POLYMER COLLOIDS

POLYMER COLLOIDS

Edited by

R. BUSCALL

*Corporate Bioscience and Colloid Laboratory, ICI PLC,
Runcorn, UK*

T. CORNER

International Paints, Felling, Gateshead, UK

and

J. F. STAGEMAN

Agricultural Division, ICI PLC, Billingham, UK

ELSEVIER APPLIED SCIENCE PUBLISHERS
LONDON AND NEW YORK

ELSEVIER APPLIED SCIENCE PUBLISHERS LTD
Crown House, Linton Road, Barking, Essex IG11 8JU, England

Sole Distributor in the USA and Canada
ELSEVIER SCIENCE PUBLISHING CO., INC.
52 Vanderbilt Avenue, New York, NY 10017, USA

British Library Cataloguing in Publication Data

Polymer colloids.
 1. Colloids   2. Polymers and polymerization
 I. Buscall, R.   II. Corner, T.
 III. Stageman, J. F.
 541.3′45   QD549

ISBN 0-85334-312-8

WITH 20 TABLES AND 89 ILLUSTRATIONS

Printed in Northern Ireland by The Universities Press (Belfast) Ltd

Preface

When we first started to plan this book we asked ourselves two questions, what sort of book would we have liked available when we first encountered polymeric colloids and what sort of book would we like on our shelves now? A discussion of these questions caused us to realise that polymer particles are of interest to a wide range of scientists and technologists with very different backgrounds and preoccupations. Thus polymer colloids are not only the domain of the polymer chemist and the colloid scientist, the paint technologist and the plastics manufacturer, they are also of concern to the condensed matter physicist and the rheologist to whom they represent useful model systems. We therefore decided to attempt to produce a book that would (a) serve as an accessible but non-trivial introduction to the field, and (b) serve as a source of information for the expert in an area who wishes to become more familiar with some other aspect of the subject. The range of subject matter to be covered leads inevitably to a book with a multiplicity of authors; it was felt, however, that what might be lost in terms of uniformity of style would be more than compensated for by the authority of the individual contributions.

The book is largely concerned with fundamentals although three of the chapters at least, those by Morice Thompson, Derek Rance and David Blackley, do contain substantial material of a more applied nature. As such they serve to show how the more fundamental considerations apply in practice. The first chapter on preparative

v

methods is a condensed but comprehensive survey of the methods that can be used to make colloidal polymer particles. The development of new methods has occurred largely in industry and this chapter describes not only those methods that are well documented in the scientific literature but also many that are only familiar to those with access to the patent literature. The second chapter examines in detail what is arguably the best understood of these methods, aqueous emulsion polymerisation. The third and fourth chapters are concerned with the adsorption of low and high molecular weight species at the polymer/liquid interface, a topic essential to an understanding of all aspects of the behaviour of polymer particles. The fifth chapter deals with the equally fundamental matter of particle stability, and to the best of our knowledge is the only recent review of this subject to give equal weight to both electrostatic and steric stabilisation. The sixth chapter deals with the rheology of latices, a subject of considerable practical importance. It is also an example of an area where research on latex systems has been of more widespread interest since the general area of dispersion rheology has benefited considerably from careful experimental work on polymer colloids. It would have been remiss of us to have produced a book on polymer colloids without considering natural latex; natural and synthetic rubber latices are compared and contrasted in Chapter 7. The final chapter is concerned with PVC, a polymer of great practical importance which is also unusual in many respects. PVC is one of few common polymers insoluble in its own monomer and this allows its preparation by novel routes. The PVC/VC system is also one of rather few non-aqueous disperse systems whose stability is influenced by electrostatic forces. The stability of PVC particles in vinyl chloride monomer has been the subject of recent research and an up-to-date account of this interesting topic is given in this chapter.

We are sad to report that Morice Thompson died suddenly during the course of the preparation of this book. He will be greatly missed by all his friends and colleagues in ICI and throughout the scientific world. To our mind Morice had a unique knowledge and understanding of the chemistry of polymer particles and we are glad that this book has allowed him to share this knowledge to some extent. Morice's chapter was left to us in draft form and we have undertaken some rewriting and revision as a result. The editors are thus responsible for any inaccuracies or defects that there may be in this contribution.

It remains only to thank the authors for their excellent contributions, the publisher for making our task a relatively painless one, and the management of Imperial Chemical Industries PLC for encouraging us in this project.

R. Buscall
T. Corner
J. F. Stageman

Contents

List of Contributors

D. C. BLACKLEY
  London School of Polymer Technology, The Polytechnic of North London, Holloway, London N7 8DB, UK

R. BUSCALL
  Corporate Bioscience and Colloid Laboratory, Imperial Chemical Industries PLC, PO Box 11, The Heath, Runcorn, Cheshire WA7 4QE, UK

R. O. JAMES
  Emulsion Research Division, Research Laboratories, Eastman Kodak Company, 1669 Lake Avenue, Rochester, New York 14650, USA

I. M. KRIEGER
  Department of Chemistry and Center for Coatings, Adhesives and Sealants Research, Case Western Reserve University, Cleveland, Ohio 44106, USA

R. H. OTTEWILL
  Department of Physical Chemistry, School of Chemistry, University of Bristol, Cantock's Close, Bristol BS8 1TS, UK

GARY W. POEHLEIN

School of Chemical Engineering, Georgia Institute of Technology, Atlanta, Georgia 30332, USA

D. G. RANCE

Petrochemicals and Plastics Division, Imperial Chemical Industries PLC, PO Box 90, Wilton, Middlesborough, Cleveland TS6 8JE, UK

THARWAT F. TADROS

Plant Protection Division, Imperial Chemical Industries PLC, Jealott's Hill Research Station, Bracknell, Berkshire RG12 6EY, UK

The late M. W. THOMPSON

Research Department, Paints Division, Imperial Chemical Industries PLC, Wexham Road, Slough SL2 5DS, UK

Chapter 1

Types of Polymerisation

The late M. W. THOMPSON

Paints Division, ICI PLC, Slough, UK

1. INTRODUCTION

This chapter represents a brief survey of the methods that can be used to prepare polymer dispersions, a summary table of which is given in the Appendix. The formation of fine polymer particles has become an important area of scientific study largely because of its industrial importance, the total weight of colloidal and particulate polymer produced worldwide now running into hundreds of thousands of tonnes per annum. Originally the impetus for producing polymers in the form of fine particles arose from a desire to find polymerisation processes which (a) could be run on a large scale without overheating, and (b) produced polymer in a physical form suitable for subsequent processing. Latterly polymer dispersions have found application in many areas of technology, notably in the adhesives, surface coatings and sealants areas where their particular physico-chemical properties can be exploited to good effect.

Nearly all of the methods used to produce polymer dispersions involve precipitation of polymer from solution at some stage in the process. If colloidal or microscopic particles are to form rather than ill-defined agglomerates, then there has to be present some means of conferring stability on the growing particles which prevent their uncontrolled aggregation or flocculation. In certain forms of polymerisation carried out in water, for example aqueous emulsion polymerisation, particle stabilisation can be electrostatic in origin, an electrostatic repulsion between the particles being generated by the presence of ionising groups at the particle/medium interface. Electrostatic stabilisation is, however, sensitive to agitation and shear and to the presence of divalent and multivalent ions in trace amounts, and so has limitations as a result; in particular it can be difficult to prepare high solids dispersions if electrostatic repulsion alone is relied on. During the development of aqueous emulsion polymerisations it was soon found that stability could be enhanced by the addition of a soluble polymer such as a polysaccharide or poly(vinyl alcohol). Such materials are now known to adsorb on the polymer particles as they form and confer on them a measure of steric stability. The last 20 years or so have seen the increasing use of carefully designed amphipathic graft copolymers as steric stabilisers in both aqueous and non-aqueous media.[1] The availability of these materials has allowed the preparation of polymer dispersions of both addition and condensation polymers in a variety of media ranging from water at one extreme to liquefied inert gas at the other.[2,3] The development of a particular polymerisation method has invariably required a blend of polymer chemistry, polymer physics and colloid science since the polymerisation reaction, the means of stabilisation and the particle size, structure and surface chemistry are always highly interrelated.

The range of methods now available to produce polymer dispersions is wide and at first sight diverse; however, as has been noted above, polymer is almost always precipitated from solution at some point, either from solution in an inert diluent or from solution in its own monomer. The various methods can be grouped or classified to some extent according to the state of solution and solubility of the various reactants (i.e. monomer, initiator, etc.), at the start of the reaction. Polymerisations in which the reactants and excipients are completely dissolved in the solvent or monomer so that the starting mixture is entirely homogeneous are referred to as *dispersion polymerisations*. Those where the initiator is soluble but the monomer is only sparingly so, so that the bulk of it exists as a separate phase consisting of suspended droplets, are referred to as *emulsion polymerisations*. If the monomer is insoluble in the diluent medium and the initiator also, so that it resides in the monomer droplets rather than in the continuous phase, then the process is a *suspension* or *microbulk* polymerisation.

Of the various types of polymerisation, aqueous emulsion polymerisation has probably been studied in the greatest detail, particularly in respect to its mechanism and kinetics. For this reason it has been given a chapter in its own right (Chapter 2); the emphasis here will consequently be more on dispersion and suspension polymerisation. However, before a discussion of these is embarked upon, certain salient features of bulk and solution polymerisation that will be of use later will be highlighted briefly.

2. BULK POLYMERISATION

The formation of polymers in the absence of solvent or diluent is difficult to control. However, in many commercial operations it is of prime importance to obtain the polymer in a particular physical form so as to facilitate downstream processing. Thus many polymerisation processes are devised so as to obtain polymer in powder, bead, chip or porous form and sometimes bulk polymerisation offers advantages in this respect. There are, however, problems of viscosity, heat evolution, mechanical handling and the control of molecular weight associated with bulk polymerisations. Some of the problems can be overcome by means of suspension or 'microbulk' polymerisation where, in essence, a bulk reaction is carried out in a small droplet dispersed in a suitable heat transfer medium. As a preliminary to a discussion of this and other types of two-phase polymerisation, some mention will be made of one or two of the more salient features of bulk polymerisation.

2.1. Radical-Induced Addition Polymerisation in Bulk

The control of heat evolution is a particular problem in these polymerisations since monomers such as vinyl chloride and ethylene have high heats of polymerisation. The exotherm accompanying the polymerisation can also be accompanied by a rapid increase in viscosity as the molecular weight rises; this, by reducing the mobility of growing chains, retards termination and transfer processes leaving monomer as the only mobile species. The result is a large increase in the effective propagation rate known as the gel effect or Tromsdorff effect. Monomers such as acrylonitrile exhibit this autocatalytic effect to such a degree[4] that explosions can occur. The Tromsdorff effect is important in many two-phase processes.

The bulk polymerisation of vinyl chloride is of particular interest because poly(vinyl chloride) (PVC) is insoluble in its own monomer. As a result particles of PVC form during the course of the polymerisation by precipitation. The polymerisation process, which typically is carried out between 40 and 70°C using isopropyl peroxydicarbonate as the initiator, occurs in two stages. In the first stage, up to ~2% conversion typically, polymer precipitates in the form of submicron particles ($0 \cdot 1$ μm or so), these then aggregate to form clusters which grow by absorbing monomer until the conversion reaches about 80%. The size of the aggregates formed depends upon the strength of agitation which in turn influences the porosity of the final product. This type of polymerisation can be thought of as an unstabilised form of dispersion polymerisation of PVC in its own monomer. The polymerisation of PVC and the influence of stabilisers and stabilising forces will be considered in detail later in Chapter 8.

Temperature is an important variable in all polymerisation processes. The effect of temperature on bulk polymerisations can be complex, affecting as it does the rate of initiator decomposition, the mobilities of various species, and so on. Of particular importance is the location of glass transition temperature of the polymer since all species are mobile above the T_g whereas only low molecular weight species are mobile below. Models for polymerisation above and below the T_g have been discussed by Hamielec and Marten.[6] In certain cases the immobility of macrospecies below the T_g, and the ability of monomers to plasticise the polymer and so reduce the T_g, can have a controlling effect on conversion. Thus the conversion of methyl methacrylate (MMA) to poly(methyl methacrylate) (PMMA) at 70°C is limited to 92% since the T_g of 92% PMMA/8% MMA is itself 70°C. On the other hand, polymerisation rates can be enhanced below the T_g in cases where the Tromsdorff or gel effect is particularly pronounced, as it is with acrylonitrile.[3] In such cases increasing the temperature past the T_g results in a reduction of rate and of molecular weight.

Various features of bulk, emulsion and solution polymerisation have been compared and contrasted by Albright[7] and Sanghvi.[8] The latter gives, from the point of view of producing polymer as distinct from a polymer dispersion, a table of advantages and disadvantages of the various processes emphasising yield and cleanliness of the product. A similar paper by Polnis and Sharma[9] provides, in addition, useful information on heats of polymerisation and the properties of initiators. Topics such as the role of the gel effect, continuous production of polymer, the properties of monomer polymer slurries and radiation-initiated solid-state polymerisation are also discussed. Amongst other things, the good heat stability of PVC produced by bulk polymerisation is emphasised in this review and this raises an important point. The good heat stability of bulk-polymerised PVC is probably attributable to the absence of inorganic ions, surfactants and other contaminants which tend to be introduced when PVC is prepared by other routes. The effect of polymerisation additives and other excipients on the physical and chemical properties of the final polymer can be pronounced and is something to consider in all heterogeneous and two-phase polymerisations. This point will be returned to again in subsequent sections.

No more will be said of free-radical addition polymerisation in bulk here. Further details can, however, be found in Ref. 10.

2.2. Condensation Polymerisation in Bulk

The formation of condensation polymers involves the elimination of a byproduct which may be volatile. Very often the byproduct is water as it is with polyethylene terephthalate, other polyesters and the nylons. The rate of such condensation polymerisations is limited in the bulk by the rate at which the byproduct can be expelled from the reaction mixture, and this in turn depends upon how far the byproduct has to diffuse in order to reach a phase boundary where it can be expelled. For this reason, and in particular since very high temperatures are required if the mean path of migration is large, condensation polymers are frequently prepared in continuous thin-film reactors where the diffusion path is very much reduced. An alternative way of reducing the mean path of byproduct migration is to perform a dispersion or suspension polymerisation in a suitable diluent medium.

3. SOLUTION POLYMERISATION

3.1. Free Radial Polymerisation in Solution

The mechanism of free radical initiated addition polymerisation in solution is of particular relevance to dispersion and emulsion polymerisations since polymerisation starts in solution in these processes. Solution polymerisation by radical initiation proceeds according to the following general scheme:

$$I : I \rightarrow I \cdot + I \cdot \qquad\qquad \text{Initiation} \qquad\qquad (1)$$

$$I \cdot + M_n \rightarrow I[M]_n^{\cdot} \qquad\qquad \text{Propagation} \qquad\qquad (2)$$

$$I[M]_n^{\cdot} + \cdot[M]_m I \rightarrow I[M]_n [M]_m I \qquad \text{Termination by combination} \qquad (3)$$

$$I[M]_n^{\cdot} + \cdot[M]_m I \rightarrow I[M-H]_n + [M+H]_m I \qquad \text{Termination by disproportionation (H abstraction)} \quad (4)$$

$$I[M]_n^{\cdot} + Solvent \rightarrow I[M+H]_n + Solvent \cdot \quad \text{Termination by solvent} \quad (5)$$
$$I[M]_n^{\cdot} + Polymer \rightarrow I[M+H]_n + Polymer \cdot \quad \text{Termination by polymer} \quad (6)$$
$$I[M]_n^{\cdot} + RSH \rightarrow I[M+H]_n + RS \cdot \quad \text{Termination by transfer} \quad (7)$$
$$\text{agent}$$

The radicals produced in the last three steps on the solvent, polymer and transfer agent can initiate new polymer chains if they are active enough and if the monomer is readily polymerised. However, radicals induced on solvent molecules are not believed to initiate new polymer chains readily whereas radicals on polymer and transfer agent can easily propagate new chains. Initiation by polymer radicals leads to grafting and thus the formation of branched chains; transfer initiation of new chains has an important regulating effect on the molecular weight of the final polymer. The scheme given above is somewhat oversimplified but it will be sufficient for the purposes of this chapter. A detailed review of solution free radical polymerisation and the factors which influence its mechanism and kinetics has been given by Matsumoto et al.[12]

The rate of initiation[13] is given by:

$$r = R_i^{\frac{1}{2}}(M)^2[K_1(M)^2 + K_2C(M)(S) + K_3C^2(S)^2]^{-\frac{1}{2}}$$

where R_i = rate of chain initiation
M = concentration of monomer
S = concentration of solvent
C = the transfer constant, and K_1, K_2 and K_3 are composite constants involving rate constants for propagation and termination.

The rate of initiation is influenced by the nature of the initiator, temperature and the nature of the solvent. Common initiators include peroxides, hydroperoxides, azo compounds, peroxy-anion salts such as persulphates, and redox couples comprising peroxy-salts and a suitable reducing agent. Redox systems apart, initiator decomposition occurs by thermolysis and the rate of this may or may not be affected by the nature of the solvent. Solvent effects are most pronounced with peroxides.

The nature of the solvent affects most polymerisations, partly through interaction of the solvent with the polymeric radical and also by its effect on the viscosity of the monomer/solvent mixture. Solvation of the growing radical can enhance the rate of polymerisation as is illustrated by the work of Okamura and Urakawa[14] who showed that the rate of polymerisation of vinyl acetate in alcohol is increased by the addition of small amounts of water. In certain cases charge-transfer complexes can form and in such cases the rate of propagation tends to be reduced.[15]

Transfer effects in solution polymerisation have been described extensively by Young.[16] The effect of a transfer agent on the molecular weight can be quantified by a transfer constant C defined as:

$$\frac{1}{P_n} = \frac{1}{P_{n,0}} + C\frac{X}{M}$$

where P_n = number average molecular weight obtained in the *presence* of transfer agent X

$P_{n,0}$ = number average molecular weight obtained in the absence of transfer agent X

C_X = transfer constant to transfer agent X

M = monomer concentration

X = transfer agent concentration

Some examples of practical solution polymerisations are given in Table 1. The first four examples are homopolymers, the latter two copolymers, solution copolymerisation being a common route to the production

TABLE 1
Typical Solution Polymerisation Processes[14]

Monomer(s)	Solvent	Initiator	T°C	Product
Methyl methacrylate	Toluene/acetone	Benzoyl peroxide	80–100	Acrylic automotive lacquer
Vinyl acetate	Alcohol	Peroxide or azo	Reflux	Poly(vinyl alcohol) by hydrolysis
Alkyl acrylate	Ethyl acetate	Azo	Reflux	Solution adhesive
Acrylonitrile	Dimethyl formamide	Ammonium persulphate	45–55	Spun into fibres from solution
Acrylamide and acrylonitrile	Water	Ammonium persulphate	75–80	Flocculating agents for purification of water
Maleic anhydride and styrene	Acetone	Benzoyl peroxide	Reflux	Water-soluble thickener

of speciality polymers. Copolymerisation in solution and the composition of the polymers formed are described by the Alfrey–Price Q–e scheme, the parameters Q and e, respectively, describing the influence of resonance and polar interactions between the two reacting monomers on the propensity for co-addition rather than homo-addition. For example, monomers with very similar Q and e values form random copolymers; if the e values are very different, i.e. the monomers have opposing polarities so that they interact strongly, alternating copolymers are formed. An extensive tabulation of Q, e values can be found in the *Polymer Handbook*.[16] In two-phase polymerisations, such as emulsion polymerisation and dispersion polymerisation, unfavourable reactivities can be offset to some extent since the two monomers may partition to differing extents between the growing particle and the solvent. Controlled seed/monomer feed and growth procedures can also be used to influence polymer composition in these processes, examples will be given later.

3.2. Ionic Polymerisation in Solution

Anionic polymerisation in solution using alkali-metal alkyls or aryls has been used extensively to prepare polystyrene, polyisoprene and polybutadiene of controlled molecular weight. The mechanism of anionic polymerisation has been described by Richards.[17] The method can also be used to produce graft copolymers of precise composition and so can be used to prepare stabilisers for use in two-phase polymerisations. At the end of an initial homopolymerisation stage the terminal living anion is reacted with a terminator which introduces a functional group, for example a vinyl group or an epoxide group, which can then be used to grow on a new chain of different chemical composition. A review by Athey[18] gives details of the range of functional end-groups that can be introduced.

Coordination complex polymerisation, including Ziegler–Natta catalysis, is reviewed by Tait *et al.*[20] Transition-metal alkyl catalysis is of interest here since compounds such as tetra-benzyl titanium can be supported on fumed silica and in conjunction with an aluminium diethyl chloride co-catalyst can be used to polymerise alkathenes so as to produce dispersions in hydrocarbons.

3.3. Polycondensation in Solution

Certain condensation polymers can be formed in solution, for example by the reaction of acid chlorides or anhydrides with diamines in polar solvents such as dimethylacetamide or dimethylformamide. Chain breaking can, however, occur in the case of acid chlorides by attack on the solvent,[21] this leads to a reduction in molecular weight. Dispersion polymerisation may be a way of avoiding this problem.

4. PRECIPITATION POLYMERISATION IN THE ABSENCE OF DISPERSION STABILISERS

Polymerisation in solution followed by precipitation is the initial step in many processes used to make polymer dispersions. In the case of amorphous polymers precipitation usually occurs because the entropy of solution of the polymer is smaller than that of the monomer (the

TABLE 2

The Morphologies of Polymers Precipitating from Monomer/Solvent Mixtures

Monomer	Solvent	Initiation	Properties of the precipitate
Ethylene	Heptane	Ionic	Fine particles which aggregate but do not absorb monomer.
Vinyl chloride	Vinyl chloride	Radical	$0.2\ \mu$m particles which aggregate. Polymer absorbs own monomer.
Vinyl chloride	Heptane	Radical	$0.2\ \mu$m particles which aggregate but do not absorb own monomer to any great extent.
Acrylonitrile	Acrylonitrile or hexane	Radical	Particles with many trapped radicals present aggregate and then polymerise violently (explosion Tromsdorff effect).
Methyl methacrylate	Heptane or cyclohexane	Radical	Particles formed and then coarse aggregates as polymer absorbs own monomer.
Ethyl acrylate	Hexane	Radical	Fine particles which then aggregate to form molten mass of polymer/monomer mixture due to low T_g of polymer.
Vinyl acetate	Water or hexane	Radical	Fine particles and then coalescing aggregates as monomer is absorbed into the particles. No acceleration in rate is observed unless polymer is below T_g (30°C).[22]
Styrene	Heptane	Ionic	Fine particles aggregate round initiator.

dissolution process is usually endothermic unless perhaps the monomer and solvent are very polar). In the case of crystallising polymers the ability to form crystalline zones in the solid phase enhances the insolubility of the polymer. Crystallising polymers (e.g. PVC) can be insoluble in their own monomers, amorphous polymers are rarely so.

The physical form and morphology of the material that precipitates from a polymerising system and of the aggregates and accretions that form if no means of imparting colloid stability to the precipitate is present depend upon the T_g of the polymer, its solubility, the extent to which it is plasticised by its own monomer or by solvent, and so on. The characteristic morphologies shown by a number of common polymers precipitating from liquids which are miscible with their monomers are given in Table 2. The type and structure of precipitate formed have implications for the development of a controlled two-phase process for preparing stable particles since the nature of the precipitating polymer influences the mechanism and kinetics of particle formation and growth, even to the extent of determining which processes are possible. For example, an important requirement in free radical dispersion polymerisation is that the particles be capable of absorbing monomer. It is also easier to maintain colloidal stability of the growing particles throughout a process if the precipitate is amorphous rather than crystalline, and soft and swollen rather than hard.

5. POLYMERISATION AT SURFACES—THE EFFECT OF INTERFACES ON ADDITION POLYMERISATION AND THE BEHAVIOUR OF SURFACE ACTIVE MONOMERS AND RADICALS AT INTERFACES

Polymerisation at interfaces can give very high rates of polymerisation and produce polymers of very high molecular weight. This is because the adsorbed monomers are restricted to a two-dimensional layer which limits their freedom and which may also align them in positions most favourable for polymerisation. An analogous situation may also arise when monomers are polymerised in the presence of a preformed polymer with which they interact strongly so that the polymer acts as a template. An example of such a polymerisation is the polymerisation of methyl methacrylate in the presence of poly(vinyl pyrrolidone) (PVP)[23,24] where an increase in propagation rate is observed which depends upon the tacticity of the PVP. Template polymerisation of this type was first described by Bamford and co-workers,[25-27] and the subject has also been reviewed.[29,30]

There have been a number of investigations of the polymerisation of oriented monomers at interfaces on a Langmuir trough using surface-active monomers such as octadecyl methacrylate, 16-oxo-16-(p-vinylphenyl) hexadecanoic acid, octadecyloxypropyldiacrylate and similar long-chain compounds containing polymerisable double bonds.[30-38] A feature of such polymerisations is the high degree of two-dimensional crosslinking that can be obtained if a difunctional monomer such as divinyl benzene is added.

In view of the fact that preformed polymer is often present and that the interfacial area of the dispersed phase(s) can be large, it would be surprising if template and interfacial effects were not important in processes such as dispersion polymerisation and emulsion polymerisation, although to the author's knowledge it appears that their role has yet to be studied seriously in this context.

6. POLYMERISATION AS A MEANS OF PRODUCING POLYMER DISPERSIONS—GENERAL FEATURES

Dispersions of particles in liquid media are usually considered to be colloidal if the particles have sizes which lie in the range 0·01–1 μm, although this definition is a little arbitrary. The methods described in the remainder of this chapter are capable of producing particles with sizes in this range and, in many cases, dispersions with precise size distributions that can be varied over a wide range by changing the conditions.

There are four general starting points for making polymer dispersions:

(1) An emulsion of monomer in an inert liquid which produces insoluble polymer particles (as in emulsion polymerisation, suspension polymerisation).
(2) A solution of monomer which produces an insoluble polymer (dispersion polymerisation).
(3) Dispersions of solid monomers giving solid, insoluble polymer.
(4) Polymer solutions caused to precipitate by the addition of non-solvent.

In each case some form of stabilisation is required if colloidal particles are to form. Further, the method of stabilisation must be compatible with the type of polymerisation and method of initiation or catalysis. It must also be able to accommodate any morphological changes that occur during or after polymerisation, such as crystallisation of the polymer within the particles.

6.1. Mechanism of Stabilisation

Colloidal polymer particles may be stabilised by one of two mechanisms:

(1) Electrostatic repulsion arising from the presence of ionised moieties at the particle surface, i.e. *electrical double layer repulsion.*

(2) The repulsion generated between solvated layers of chain molecules at the surface of the particles, i.e. *steric stabilisation.*

Steric stabilisation can be used in all types of media; electrostatic stabilisation, on the other hand, is rarely effective in media other than water.[39] The subject of colloid stability is discussed at length in Chapter 4 and the adsorption of stabilising species at the liquid/particle interface is considered in Chapter 3. Thus only specific points which relate to particle formation will be made here. An important difference between the two mechanisms lies in their effect on the approach of small molecules to the particle surface, a steric barrier will not normally inhibit the approach of small molecules since these can easily penetrate the adsorbed layer. In the case of a charged particle, however, the close approach of all like-charged species is inhibited regardless of size. This observation has important mechanistic connotations.

6.1.1. Electrical Double Layer Stabilisation

This is primarily of importance in the context of aqueous emulsion polymerisation and this topic is considered in detail in Chapter 2. The necessary electrical charge at the particle surface usually arises in part from the presence of ionic chain end-groups which in turn arise from the use of an ionic initiator such as a persulphate. The stability can, however, be very much enhanced and finer dispersions obtained by the addition of an ionic surfactant, for example sodium dodecyl benzene sulphonate. Adsorbing polyelectrolytes can also be used.[40,41]

6.1.2. Steric Stabilisation

Steric stabilisation can be used in all liquid media for which soluble stabilising macromolecules can be found, in practice this appears to mean any non-metallic organic or inorganic liquid. Thus steric stabilisation has been used to prepare dispersions of addition polymers, condensation polymers, ring-opening polymers and inorganic polymers (silica) in a wide variety of aqueous and non-aqueous media. It is apparent then that steric stabilisation is a much more flexible adjunct to particle preparation than is electrical double layer stabilisation.

6.1.3. Preparation of Steric Stabilisers

The basic requirements for steric stabilisation are:

(1) The particle surface should be covered by a contiguous layer of solvated chain molecules.
(2) The anchoring of the chains should be good, ideally they should be either chemically or irreversibly adsorbed.

It is rarely the case that a homopolymer will fulfil both of these requirements and so stabilising molecules are usually block or graft copolymers which contain distinct anchor and stabiliser chains. The various types of stabiliser that can be used in the preparation of sterically stabilised dispersions of polymers formed by addition, ring-opening and condensation polymerisation have been described by Walbridge.[42] The different types can be classified as follows.

6.1.3.1. Stabiliser precursors. When a low molecular weight polymer which possesses reactive groupings and which is also soluble in the dispersion medium is added to a dispersion/emulsion polymerisation, then it is highly likely that graft copolymers will be produced during the polymerisation. Moreover, these *in-situ* generated graft copolymers are quite often found to be very efficient steric stabilisers. Hence such polymers were termed stabiliser precursors by Waite.[43] The functional groups must be copolymerisable by the mode of polymerisation utilised and ideally the stabiliser precursor should possess a single functional group at one end only. However, a random distribution of the functional groups is satisfactory provided that on average only one functional group is present per chain. For addition polymerisations the stabiliser precursor should possess a polymerisable unsaturation such as vinyl or acrylic. Such compounds have been termed macromonomers or 'macromers' and anionic polymerisation methods by which they

can be synthesised have been studied extensively by Milkovich[44] and
Dawkins and Taylor.[45] Whilst such methods in principle offer a greater
chance of preparing 100% functionalised polymers, it is doubtful
whether the narrow molecular weight distribution of the stabiliser
precursor offers any advantage over materials possessing broader
molecular weight distributions. Also, since Corner[46] has recently
shown that free radical polymerisation methods carried out under
carefully controlled conditions can in fact yield stabiliser precursors of
very high functionality, it would appear that free radical methods offer
the greatest versatility in terms of polymer type.

6.1.3.2. Preformed graft copolymers. Amphipathic graft copolymers
are widely used for stabilising colloidal particles in all types of media
and there are many methods available for their synthesis. However,
the most useful type of graft copolymer for dispersion polymerisations
is that to which has been given the trivial name 'comb graft'.[47] These
graft copolymers are composed of a single insoluble chain (the backbone) to which are attached several pendant chains of soluble polymer
(the teeth) in a comb-like configuration and they are readily prepared
by the copolymerisation of a stabiliser precursor with a second
monomer which is typically a simple acrylate or methacrylate.[42] 'Teeth'
of poly(12-hydroxystearic acid) and poly(lauryl methacrylate) are
commonly used in aliphatic hydrocarbons, poly(methyl methacrylate)
in ketones and poly(N-vinyl pyrrolidone) in aqueous media. Combs
with poly(oxyethylene-glycol monomethyl ether) side-chains can also
be utilised in aqueous media and recently such combs have been
further developed for use in both non-ionic dispersion and emulsion
polymerisation.[48,49]

6.1.3.3. Block copolymers. Preformed, amphipathic AB and ABA
block copolymers can also be successfully used to stabilise dispersions
of polyolefins and polyvinyl aromatics. These types of copolymer are
best prepared by anionic polymerisation methods so as to achieve
optimum control of block lengths and composition.[2,45,50]

6.1.3.4. Amphipathic precursors. Stabiliser precursors which are
themselves amphipathic in nature have also been used to stabilise
dispersions. The terminal, polymerisable group can be located either at
the hydrophilic end[51] or at the hydrophobic end,[52] the former location
giving the most efficient stabilisation.

7. DISPERSION POLYMERISATION BY RADICAL INITIATION OF VINYL AND ACRYLIC MONOMERS

Submicron, particle-size polymer dispersions can be obtained by micro-bulk (i.e. microbead), by precipitation and by the various types of emulsion polymerisation process. Also, although the particles produced are not strictly of colloidal dimensions, suspension or bead polymerisation should be considered since it is a method which is widely used to prepare beads of poly(methyl methacrylate) for subsequent use as moulding materials, etc. In this polymerisation process a dispersion of monomer droplets is produced by high-speed stirring using small amounts of a dispersing agent such as poly(acrylic acid). When more efficient dispersing agents are utilised it is possible to produce finer monomer droplets (hence microbead processes) which can yield colloidal-size polymer particles (see Chapter 8).

7.1. The Effect of the Locus of Initiation on the Mechanism of Polymerisation

The locus (or loci) of the initiation process during a polymerisation carried out in dispersion can have a significant effect(s) on the overall process. Where the initiation occurs solely in the monomer droplets, the final particle size and particle size distribution are largely those of the initial monomer droplet size and droplet size distribution produced during the emulsification process.

When the initiator is located exclusively in the solution or continuous phase, the polymerisation process proceeds almost entirely in solution until the polymer molecules grow to a size at which precipitation occurs. Provided that an efficient stabiliser is present, the precipitated polymer aggregates to form microparticles of colloidal dimensions. Of course it is possible for initiation to occur concurrently in both the continuous phase and within the particles. In such cases polymerisation occurs both in solution and in the particles which consequently grow by both absorption/polymerisation of monomer and adsorption of precipitated polymer.

Finally it is possible for the initiation and stabilisation mechanisms to be coupled. For example, the use of persulphates and azo-methoxypoly(oxyethylene-glycol monomethyl ether) derivatives[53] gives rise to the initiation of polymeric chains as well as providing stabilisation by ionic groups and steric barriers, respectively. This happens because unless considerable chain transfer or disproportionation reactions occur during the polymerisation, most of the polymer molecules produced will be charged at one or both ends,[54] or will be AB or ABA block copolymer molecules.[53]

7.2. The Effect of Stabilisation on the Kinetics of Polymerisation

The major effect on the kinetics of polymerisation in those cases where the polymer being produced is maintained in a colloidal state by suitable stabilisation are as follows. Firstly, the viscosity of the continuous phase generally remains low and maintains high heat dissipation characteristics. These factors, coupled with the very large surface area of the polymer particles and hence a high surface/bulk ratio, ensure that the heat of polymerisation is efficiently removed. As a result, overheating problems are less frequent and greater control of fast reactions is obtained. Secondly, the maintenance of a large surface area aids the ready absorption of monomer and initiating species, and so reduces the effects of diffusion processes on polymerisation rate. These effects have been discussed by Fitch and Kamath[55,56] and are illustrated in Fig. 1 where the rates of polymerisation for solution, precipitation and dispersion polymerisations are compared.

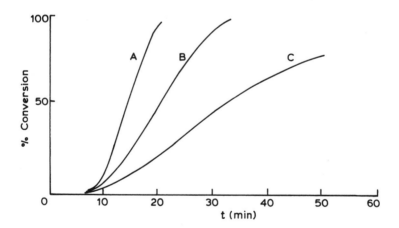

Fig. 1. The effect of stabiliser on the kinetics of polymerisation. A—Dispersion polymerisation; B—precipitation polymerisation; C—solution polymerisation. All obtained for the free radical polymerisation of methyl methacrylate in dodecane at 80°C (see Ref. 57).

7.3. Types of Addition Polymerisation in Aqueous Media

7.3.1. Microbulk

In general, microbulk polymerisations, i.e. bead or suspension polymerisation, do not yield polymer particles of colloidal dimensions. However, it is possible to make use of specially designed dispersants in the preparation of submicron-sized polymer particles, hence microbead polymerisations. For example, if methyl methacrylate is emulsified in water using a 'comb' type graft copolymer possessing poly(oxyethylene-glycol monomethyl ether) side-chains and a backbone of a methyl methacrylate/methacrylic acid copolymer in which some of the methacrylic acid units have been esterified by glycidyl methacrylate, then a stable emulsion is produced of colloidal dimensions. Moreover, when the methyl methacrylate is polymerised within the droplets this graft copolymer successfully maintains the integrity of the individual particles. This phenomenon is due entirely to the presence of the pendant, copolymerisable unsaturation attached to the graft copolymer which ensures that the stabilising copolymer molecules are covalently bound to the particles and hence are able to resist displacement/desorption as polymerisation takes place in the monomer droplets. Polymerisations carried out in an identical manner but in which the methacrylic acid units had not been esterified with glycidyl methacrylate resulted in gross agglomeration of the droplets as the graft copolymer molecules were desorbed. This type of process quite often results in the formation of significant amounts of gel within the particles. Nevertheless, it is a useful process which has been extended to systems in which preformed resins are dissolved in the acrylic monomer. On polymerisation, composite materials are thereby produced.[58,59]

7.3.2. Aqueous Emulsion Polymerisation

This type of polymerisation is discussed in detail in Chapter 2.

7.3.3. Sterically Stabilised Aqueous Dispersion Polymerisation

The characteristics of, and the differences between, the various types of non-aqueous polymerisation are described in a book on dispersion polymerisation in non-aqueous media edited by Barrett.[1] Recently, aqueous analogues of the non-aqueous dispersion polymerisations described therein have been developed.[48,60] Unlike in aqueous emulsion polymerisations where the monomers are emulsified to give an oil/water emulsion in which the aqueous phase contains the dissolved initiator, aqueous dispersion polymerisations begin with a homogeneous reaction mixture containing uncharged initiators such as azobisisobutyronitrile. The homogeneity in the initial stages is achieved by the use of between 30 and 70% v/v of a simple alcohol such as methanol. The presence of the alcohol facilitates the use of a wider range of initiator types than in conventional aqueous emulsion polymerisation, acts as a transfer agent to modify the molecular weight of the polymer being produced without using normal transfer agents, aids in temperature control and heat removal if polymerisations are carried out at reflux, and enables dispersions of high volume fraction to be obtained by removal of the alcohol when polymerisation is complete. Typical stabilisers which are used are poly(oxyethylene-glycol monomethyl ether) acrylate or methacrylate either as stabiliser precursors or as preformed 'combs' in which the backbones are designed to possess some degree of water solubility. For example, a copolymer of poly(oxyethyleneglycol monomethyl ether) with hydroxyethyl methacrylate can be used to prepare latices of methyl methacrylate/butyl acrylate copolymers. Although it is possible to prepare polymer dispersions by a batch, i.e. a one-shot process, it is preferable to use a semi-batch, i.e. a seed and feed process, since this ensures narrower particle size distributions and enables dispersions with higher solids contents to be prepared. The seed stage typically consists of all of the diluent alcohol and water, 10% of the monomers and a proportion of the stabiliser, the initiator and any extra transfer agent if required. These components are reacted at 60–80°C under a solvent recycle system and the remaining monomers, stabiliser, initiator and transfer agent are fed into the reactor via the cold solvent return. In this manner the feed is diluted and problems of seeding are vastly diminished. Usually the dispersions are made at 45–50% w/w solids and then vacuum-stripped to remove the alcohol so as to yield dispersions with solids contents of 65–70% w/w.

To date, a number of investigations of the mechanism of these aqueous dispersion polymerisations have been carried out.[61,62] All of the results obtained indicate that the mechanism is analogous to that occurring in non-aqueous dispersion polymerisations. However, one notable feature of aqueous dispersion polymerisations is the marked dependence of the stability of the dispersions during the polymerisation on the glass transition temperature (T_g) of the polymer being produced. It has been found that dispersions flocculate during preparation if the T_g of the polymer is near or above the temperature of the polymerisation. For example, during the polymerisation of methyl methacrylate in methanol/water mixtures at 75°C, gross flocculation occurs at between 10 and 15% conversion. In an equivalent experiment using a mixture of methyl methacrylate/butyl acrylate in order to reduce the T_g, a stable latex is produced. It has been found that the flocculation problem can be overcome by lowering the molecular weight of the polymer being produced, by the addition of a small amount of a good solvent for the polymer being formed (e.g. toluene or ethyl acetate) or by the addition of a small amount of styrene. Also, the flocculation problem does not occur in the homopolymerisation of styrene. This is thought to be due to the greater extent of plasticisation of polystyrene by its own monomer.

The importance of polymerisation temperature relative to the T_g of the polymer being formed may in part be due to its effect on the ability of graft copolymer molecules to transfer from micelles to the surface of growing particles since in aqueous media the critical micelle concentration of typical graft copolymers can be several orders of magnitude smaller than critical micelle concentrations in non-aqueous media. A further factor may be the greater immobility of graft copolymer molecules on the surface of particles with a high T_g. Certainly in non-aqueous dispersion polymerisations reference has been made to 'hard' and 'soft' anchoring with respect to the nature of the anchoring component of the 'comb' stabilisers.[64]

7.4. Sterically Stabilised Addition Polymerisation in Organic Liquids

Non-aqueous dispersion polymerisation has been the subject of many publications.[1,65–70] In general, the bulk of the reported work has been carried out in hydrocarbon media using stabiliser precursors, preformed graft copolymers or AB block copolymers as the means of providing the necessary stabilisation. Although the initial polymerisation occurs in solution, precipitation occurs quickly as the polymer molecules grow to a size beyond which they are no longer soluble. As the monomer(s) is consumed, the polarity of the medium decreases and therefore precipitation occurs at a lower average degree of polymerisation. The precise pathway which is then followed is dependent upon the characteristics of the monomer being polymerised and the polymer being produced as well as the abilities of the monomer and diluent to swell the precipitating polymer.

The particle size of dispersion polymers produced in this manner is between 0·05 and 10·0 μm. Usually the particle size distributions are unimodal and very narrow but it is possible to produce polymodal distributions if required. Powders prepared from dispersion polymers are oleophilic due to the presence of the stabilising chains on the particle's surfaces and, provided that the T_g of the polymer is high, they can be either readily redispersed in hydrocarbon media or dissolved in suitable solvents.[71]

7.4.1. From Soluble Monomers Giving Precipitating Polymer Growing by Accretion

This type of process is exemplified by the polymerisation of vinyl chloride (VC) in heptane/hexane mixtures using as the stabiliser precursor poly(lauryl methacrylate) possessing pendant allyl groups.[72] Initially, polymerisation occurs in solution as both homopolymer and in-situ generated graft copolymer are formed. However, eventually precipitation occurs followed by aggregation as copolymer-stabilised particles are produced. If no additional stabiliser precursor is added at this stage, flocculation occurs quite rapidly as the stabilising chains are buried by the polymer which is continuously being formed, precipitated and then 'mopped-up' by the stabilised particles. The kinetics of the polymerisation remain those of a solution polymerisation throughout and there is no evidence of any auto-acceleration.[55] Finally, despite many attempts using graft copolymers of widely divergent compositions, it has not been found possible to successfully carry out the dispersion polymerisation of VC in aliphatic hydrocarbons using preformed stabilisers.

7.4.2. From Soluble Monomers Giving Precipitating Polymer which then Absorbs its own Monomer

The dispersion polymerisation of methyl methacrylate in aliphatic hydrocarbons in the presence of a graft copolymer stabiliser is an excellent example of this type of process.[1,69,73] Typically the polymerisations are carried out at 80°C, i.e. some 25°C below the T_g of the poly(methyl methacrylate) but probably at about the 'environmental T_g' of the poly(methyl methacrylate) present in the reaction which is undoubtedly plasticised by its own monomer, a statement which has been confirmed by the kinetic studies of Barrett.[57] These studies showed that the polymerisation rate in the dispersion process is some five times greater than in the solution case. Also, during a large part of the dispersion polymerisation, polymerisation is essentially that of a microbulk type during which high molecular weight polymer is produced very quickly compared to a normal solution polymerisation. Poly(methyl methacrylate) prepared by this process contains little, if any, gel and can be of very high molecular weight. Moreover, dispersions with very high solids contents can be prepared (see Section 11.1), either in batch or in continuous processes (see Section 11.6) via the control of the particle size and particle size distribution. Finally, the feeding of materials such as resins, reactive materials and dyestuffs into dispersion polymerisations via the monomer feed enables many interesting, sterically stabilised polymer dispersions to be produced (see Section 11.4).

7.4.3. Emulsion Polymerisation of Insoluble Monomers in Organic Liquids

This class of dispersion polymerisations has not been explored to any great extent. In principle it divides into two classes:

(1) Insoluble vinyl or acrylic monomers dispersed in hydrocarbons.
(2) Aqueous solutions of vinyl or acrylic monomers emulsified into an organic liquid which does not dissolve the monomer.

The first class of reactions is exemplified by the polymerisation of β-ureidoethyl methacrylate dispersed in an aliphatic hydrocarbon which is a non-solvent.[74] The copolymer stabiliser has to act both as an emulsifier and as an efficient stabiliser for the polymerising droplets and the final polymer particles. This multiple function is discussed further in Section 9.2 opposite the dispersion polymerisation of condensation polymers.

There are a few examples[75] of polymerisations which fall into the second class although the avoidance of the high viscosities encountered in the straight aqueous solution polymerisation should be a considerable advantage of this approach.

8. DISPERSIONS BY ADDITION POLYMERISATION IN ORGANIC LIQUIDS USING IONIC INITIATION

Steric stabilisation of particles in aliphatic hydrocarbons is a particularly appropriate method for the formation of polyolefins, poly(styrene), poly(vinyl pyridine) and poly(*tert*-butyl methacrylate). An introduction to the early work in this field may be found in the literature.[1,76-80] The success of the route relies heavily on the ease by which aliphatic hydrocarbons may be freed of protic impurities. The stabilising polymer may be selected from those which are soluble in the dispersion medium, commonly such materials as poly(octene), poly(isoprene), poly(butadiene), poly(tert-butyl styrene) and poly(dimethyl siloxane). Attachment of this stabilising moiety to an anchor component is conveniently achieved by terminating the 'living' polymer anion with a molecule which contains a subsequently polymerisable group. Examples of such terminators are epibromohydrin and allyl bromide which would leave a terminal epoxide group and an unsaturated residue, respectively. The use of a large excess of terminator avoids unwanted addition reactions with the functional groups. Ziegler–Natta catalysts have been used to produce poly(ethylene) dispersions in pentane stabilised by a block copolymer of ethylene/propylene, and dispersions of poly(propylene) stabilised by poly(octene-1).[78,79]

8.1. AB Block Copolymer Dispersions

Fine dispersions of block copolymers may be made by the sequential anionic polymerisation of two monomers which form soluble and insoluble polymers, respectively. An example of this approach is the addition of styrene to a solution of 'living' poly(tert-butyl styrene) anions which yields a stable dispersion of polystyrene when reacted to completion.[82]

8.2. ABC Block Copolymer Dispersions

If a third monomer such as isoprene is inserted between the polystyrene and the poly(tert-butyl styrene) blocks, a fine dispersion of an ABC block copolymer in aliphatic hydrocarbon can be achieved. In this case the polystyrene blocks form the particle cores, and the soluble polyisoprene and poly(tert-butyl styrene) form the stabilising chains, as shown in Fig. 2.

8.3. Polymer Dispersions Stabilised by AB or ABC Block Copolymers

Previous work by ICI[83] had given n-phenyl nylon-1 dispersions in aliphatic hydrocarbons by the polymerisation of phenyl isocyanate in the presence of a block copolymer. However, a much more extensive area was opened up by Everett and Stageman[2] and Dawkins and Taylor[45] with the preparation of precisely controlled polystyrene and

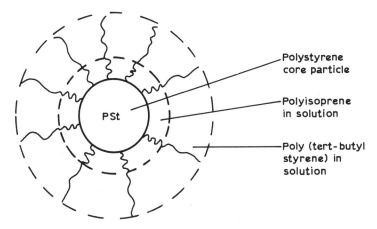

Fig. 2. Morphology of a latex particle stabilised by an ABC block copolymer dispersant.

poly(methyl methacrylate) dispersions using block copolymers with narrow molecular weight distributions as stabilisers. Dawkins and Taylor prepared AB block copolymers with poly(dimethylsiloxane) (PDMS) blocks of $\bar{M}_n$ 11 200 and poly(styrene) blocks of $\bar{M}_n$ 8800 and used these copolymers in the anionic polymerisation of styrene and the free radical polymerisation of MMA. Dispersions with narrow particle size distributions were obtained in which the PDMS blocks occupied the same surface area regardless of the particle size, indicating complete surface coverage. Everett and Stageman[2] prepared both poly(acrylonitrile) and PMMA dispersions in liquid hydrocarbons and in liquid xenon using ABA PDMS-co-PSt-co-PDMS block copolymers.

Organic fillers have been produced by anionic polymerisation in aliphatic hydrocarbons using block copolymers as dispersing agents.[84] The level of dispersing agent was used to control the size of the filler particles and the termination reaction to control the nature of the functional groups attached to the particle surface. An example of this approach was the preparation of 500-Å particles of polystyrene, which could be crosslinked by the addition of a small proportion of divinyl benzene to the monomer, using a poly(isoprene)-co-poly(styrene) dispersant. Termination of the living polymeric anions using thiuram disulphide introduced groups capable of vulcanising surrounding polybutadiene rubber whereas termination with diisocyanate yielded surface groups to which dyestuff molecules could be anchored.

9. DISPERSION POLYMERISATION PROCESSES FOR CONDENSATION POLYMERS IN ORGANIC LIQUIDS

The formation of condensation polymers in dispersion in organic liquids is summarised by Barrett and Thompson.[85] In addition to the AB and ABA types of block copolymer described earlier, 'comb' type graft copolymers have also been found to be particularly useful as stabilisers. These are composed of side-chains (teeth) which are soluble in the organic medium and an insoluble backbone which will associate with the polymer particles. This association is required to withstand the chemical and morphological changes which occur during the formation of the polymer and which are more extensive than in addition polymerisations due to both the elimination of byproducts and the change in the substrate polarity as the reaction proceeds. The elimination of byproducts such as amine hydrochlorides or the use of strong solvents such as ethylene glycol can have a strongly adverse effect on the stability. The short diffusion path from the centre to the exterior of a dispersed microparticle of condensation polymer facilitates the removal of these low molecular weight byproducts and this often leads to increased rates of polymerisation at lower temperatures. Initially the processes require methods of obtaining the monomers or reactants in a finely dispersed form. Precipitation from solution is a convenient method but many of the common reactants used to make bulk condensation polymers are not readily soluble in organic solvents so that either pre-emulsification or dispersion is required. Preferably both reactants should be contained in the same phase and the inert solvent be selected such that it eliminates any undesirable interaction which may limit the molecular weight. Some of the aspects of dispersion polymerisation have been adapted for the formation of inorganic polymer particles.

9.1. Emulsions of Liquid or Molten Reactants in an Inert Liquid

The most typical example of this class of polymerisations is the emulsification of molten bishydroxy ethyl terephthalate, containing manganese acetate as catalyst, in high boiling aliphatic hydrocarbon. The initial dispersion is achieved using high shear agitation in the presence of a graft copolymer of the 'comb' type with hydrocarbon soluble side-chains attached to an insoluble backbone. When a fine emulsion has been obtained, the monomer is converted to polymer by heating at reflux until the ethylene glycol has been eliminated by azeotropic distillation. This method is capable of producing 1–10 μm polymer particles of molecular weight $\approx$15 000, using a reaction time of 1–2 h at a reflux temperature of 190–210°C.[86] The three functions of the dispersant are that it must be capable of emulsifying the monomer, maintaining it in dispersion while the byproduct is eliminated and then anchoring securely to the new polymer surface. This has important implications for the internal composition and overall molecular weight of a successful graft copolymer as illustrated in Table 3.[87]

TABLE 3
Nylon 11 Preparation using Various Dispersants[87]

Dispersant (6% on monomer)[a]	Time to emulsify	Particle size (μm)	Time of reaction	Stability Quenched	Slow cooled
PHS/St 50/50	Slow	10–80	2 h 30 min	Fluid	Badly flocculated
PHS/St/GMA 50/45/5	Slow	3–15	2 h 20 min	Fluid cenospheres	Badly flocculated
PHS/PEG-M/St 50/25/25	Rapid	1–10	1 h 40 min	Fluid cenospheres	Badly flocculated
PHS/PEG-M/St/GMA 50/25/22·5/2·5	Rapid	0·5–5	2 h	Fluid cenospheres	Fluid cenospheres

[a] PHS = poly(12-hydroxy stearic acid)/GMA adduct; PEG-M = methoxy poly(ethylene glycol) 750 methacrylate; St = styrene; GMA = glycidyl methacrylate.

Table 3 shows that the most effective dispersant/emulsifier for nylon 11 preparation was found to be poly(oxystearate methacrylate) copolymerised with methoxypolyoxyethylene-750-methacrylate, styrene and glycidyl methacrylate. Although rapid emulsification was achieved by a poly(oxystearate methacrylate)/methoxypolyoxyethylene-750-methacrylate/styrene graft copolymer, it is probable that poor anchoring of the dispersant to the forming polymer caused the subsequent flocculation. This problem was eliminated by including certain groups in the backbone of the graft copolymer which were capable of reacting into the polymer surface as the polymerisation proceeded. Similar processes with oligomer-free bishydroxyethyl terephthalate have given dispersions of PET with particle sizes in the 0·2–0·5 μm range.

A further problem is encountered in the case where the polymer is able to crystallise as the dispersion cools. It was noted by B. R. Asher when working with the author that dispersions of such polymers always flocculated in bulk whereas small samples remained stable. It was suspected then that crystallisation was disrupting the stabilisation. This was subsequently confirmed by X-ray crystallography, IR and analysis of the level of free dispersant found in the continuous phase. In the cases where flocculation had occurred, the particles were highly crystalline and there was more free dispersant in the continuous phase. Samples which had been quenched from the reaction temperature to 60°C were stable and showed much lower crystallinity. This phenomenon was observed with PET and nylons 11, 6:6, 6, etc. Copolymers or crosslinked polymers which did not crystallise were found to produce stable dispersions. The reaction of diisocyanates with dihydroxy polymers in aliphatic hydrocarbons has been used to produce polyurethane dispersions by Union Carbide.[88] The emulsification of one reactant by a dispersing agent followed by an interfacial reaction with the second reactant also gave polyurethane dispersions from which the polymer could be recovered by coagulation and filtration.

9.2. Emulsions of a Solution of Reactants in a Second Inert Liquid

In the formation of random copolymers by condensation polymerisation in dispersion, it is often convenient to dissolve the monomers in a solvent which is immiscible with the main dispersion medium. This ensures complete homogeneity of reactants and reduces the tendency of 'blocky' copolymers to form. Examples of this approach are when mixtures of nylon salts (e.g. 6:6 plus 6:9) or 11-amino undecanoic acid and 6-amino caproic acid are dissolved in ethylene glycol which is then emulsified by high shear into a high boiling aliphatic hydrocarbon using a 'double' comb graft copolymer. The monomers are subsequently converted to polymer by azeotropic removal of the condensation product and the ethylene glycol. An example of this process is given below.

9.2.1. Emulsified Reactant Solution: Nylon 11 Dispersion in Aliphatic Hydrocarbon

A laboratory reactor is charged with 11-amino undecanoic acid, ethylene glycol, aliphatic hydrocarbon (boiling range 170–210°C) and graft copolymer dispersant (PHS/PEG-M/PMMA) using 6% dispersant on polymer. After the emulsification of the hot glycol solution using a turbine stirrer, glycol and water are removed by azeotropic distillation to give a fluid dispersion of poly(undecano-amide) (nylon 11) which is quenched by rapid cooling. The particle size range of 1–5 μm is controlled by the efficiency of the emulsification process and to a lesser extent by the level of dispersant.

9.3. Precipitation from Soluble Reactants

The formation of polymer dispersions from reactants which are dissolved in a solvent are described in Ref. 90. Fast reactions of diacid chlorides and aromatic dihydroxy compounds in the presence of an acid acceptor gave precipitates which were stabilised by a copolymer of lauryl methacrylate, methyl methacrylate and methacrylic acid. Fine particle dispersions in the range 0·2–0·5 μm were obtained but the simultaneous precipitation of the acid acceptor as a crystalline salt caused some complications.

9.4. From Solid Reactants in an Inert Liquid

Submicron stable dispersions of polyamides in aliphatic hydrocarbons can be made by the milling of the solid reactants in the presence of a soluble polymer which acts as a dispersing agent.[91] An example of the method is given below.

9.4.1. Ball-Milled Solid Reactant: Preparation of Nylon 11

11-Amino undecanoic acid, aliphatic hydrocarbon (boiling range 170–210°C) and graft copolymer dispersant (PHS/EA/MAA) were charged into a ball mill and ground for a total of 230 h. The dispersion was inspected at approximately 50-h intervals up to 150 h and on each occasion found to have thickened due to the increasing surface area of the solid monomer particles and the deficiency of dispersant. Aliquots of dispersant were added at each of these stages together with a further amount at the end of the milling when the dispersion was diluted to about 30% solids with further aliphatic hydrocarbon. The product was fluid and the dispersed particles had a mean diameter of about 1 μm. Conversion to polymer was achieved by heating the dispersion in a stirred reactor and removing the condensation product, water, by azeotropic distillation. Rapid 'quenching' of the product to 40°C gave a stable, fluid dispersion of amorphous nylon 11. Slow cooling gave a solid crumbly mass which filled the reactor, probably due to dispersant rejection as the nylon particles were allowed to crystallise.

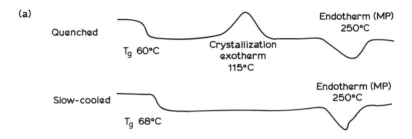

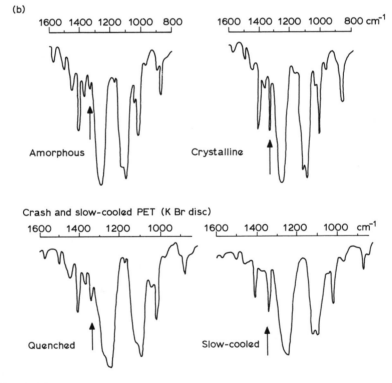

Fig. 3. (a) DSC evidence for crystallinity of PET powders. (b) IR evidence (spectra from Ref. 121).

9.4.2. Crystallisation and Flocculation Behaviour
The above example provides dramatic evidence for the important effect of crystallinity on the colloidal stability of dispersions of condensation polymers. Notwithstanding the inclusion of reactive groups (in this case methacrylic acid) into the dispersant which were capable of reacting into the growing polymer, stable dispersions were only achieved when crystallisation was prevented. Further evidence for this assertion is presented in Fig. 3.

9.5. Inorganic Polymerisation in Dispersion
The technique of emulsifying a solution of reactants in an inert liquid has been used to prepare sterically stabilised particles of poly(aluminium phosphate).[92,93] Particular care must be taken to design a dispersant that contains suitable groups, such as silanes, which are capable of reacting with the inorganic polymer as it is formed. Stable dispersions of 1–10 μm diameter particles can be made and these have oleophilic surfaces.

Stable, non-aqueous dispersions of silica can be made from carefully dried silica powders, provided great attention is paid to the mode of anchoring of the stabilising soluble polymer. In one example[94] the dry silica surface was first reacted with methyl silyl trichloride, to which could be grafted 'living' polystyrene anions in a hydrocarbon medium. In a second example[95] a graft copolymer was synthesised by the reaction of poly(isobutylene succinic anhydride) with tetraethylene pentamine. This graft copolymer adsorbed strongly onto the (dry) silica surface from an alcohol/toluene mixture and after removal of the alcohol, stable dispersions in toluene were obtained.

10. DISPERSIONS BY RING-OPENING POLYMERISATION IN ORGANIC LIQUIDS

Owing to the insolubility or the reactivity of many polymers made by ring-opening polymerisation, it is difficult to make preformed graft copolymers which act as efficient stabilisers. It is preferable, therefore, to make the graft copolymer from a polymeric precursor which contains a reactive group which can either initiate or participate in the main polymerisation.

10.1. Cyclic Ethers

The simplest stabiliser precursor, for an epoxide or oxetane polymerisation, would be a polymeric chain which contains on average one epoxide group such as lauryl methacrylate mixed with glycidyl methacrylate in the ratio 99:1. Precursor polymers which contain anhydride, carboxyl and hydroxyl groups can also be used as part of an initiating system with boron trifluoride etherate.[90] Using this approach dispersions of poly(oxyethylene), poly(styrene oxide) and poly[3:3 bis(chloromethyl) oxetane] have been made in aliphatic hydrocarbons. Aliphatic soluble glycidic compounds such as glycidyl stearate may be polymerised by boron trifluoride and then a dispersion prepared by the addition of an epoxide compound which polymerises to give an aliphatic insoluble polymer such as phenyl glycide ether.

10.2. Lactones

Polymers from β-caprolactone or β-propiolactone may also be obtained as fine particle dispersions in aliphatic hydrocarbon using poly(lauryl methacrylate) as stabiliser and dibutyl zinc as a catalyst.[96]

10.3. Acetals

Polyacetals from trioxane have been prepared using poly(lauryl methacrylate)-co-glycidyl methacrylate using the boron trifluoride etherate catalyst mentioned in Section 10.1.[90] Poly(formaldehyde) dispersions may be made in a similar manner.

10.4. Lactams

The formation of polylactams in aliphatic hydrocarbons containing dissolved rubber has been reported.[97] A copolymer of lauryl methacrylate and N-methacryloylcaprolactam was used as an activating stabiliser in the sodium-initiated polymerisation of caprolactam.[90]

11. SPECIAL TYPES OF DISPERSION POLYMERISATION

Dispersion polymerisation of acrylic monomers in aliphatic hydrocarbons using steric stabilisation has a number of attractive features which can be exploited technologically. These may be summarised as:

(1) The fine particulate state.
(2) The oleophilic character of the surface to which specially tailored stabiliser molecules can be permanently attached.

(3) The rapid rate of polymerisation and the very high level of conversion to high molecular weight polymer.

(4) The ability to control the composition and physical properties of the disperse phase polymer by altering the feed regime employed. In the former case, complete change of monomer has given heterogeneous particles which vary in structure from a 'currant-bun' type morphology at one extreme through a 'core-shell' type morphology at the other. In the latter case changes in stabiliser feed level control particle size distribution and dispersion stability whereas the introduction of bifunctional monomers gives a controllable level of crosslinking.

(5) The flexibility of being able to introduce non-polymerisable resins, dyestuffs or other chemicals into the feed.

These features are also characteristic of aqueous dispersion polymerisation (Section 7.3.3) although such precise synthetic control is more difficult to achieve.

In the case of condensation polymerisation in dispersion, an important feature is the ease with which the particles can be crosslinked internally as polymerisation proceeds; this can be done without destroying the dispersion stability since no particle growth is involved. This also allows the dispersion of pigments or filler in the monomer droplet before polymerisation has taken place.[98]

11.1. Very High Solids Addition Polymerisation in Aliphatic Hydrocarbons

In these processes, fluid dispersions of acrylic polymers and copolymers are obtained at 75–85% weight solids by using the monomer as a constituent of the continuous phase. This produces large particles but as the monomer is consumed the solvency of the medium declines, changing the precipitation conditions giving a new crop of fine particles. This type of process thus gives dispersions of a wide particle size distribution and hence a high packing fraction. Alternatively, the level of stabiliser introduced with the feed can be increased so that new crops of seed particles are formed.[99] A plant process using these principles has been in operation in ICI for over 10 years giving a fluid 76% solids acrylic copolymer dispersion in aliphatic hydrocarbon.

11.2. Heterophase Addition Polymerisation

Work by Osborne[100] has shown that two-phase polymer particles can be obtained by making a seed dispersion of PMMA in an aliphatic hydrocarbon and then feeding ethyl acrylate to produce inclusions of PEA in a PMMA matrix. The exact converse structure can be synthesised by reversing the order of the feed with the level of inclusion being controlled by the degree of crosslinking introduced into the matrix polymer.

Further work in non-aqueous dispersions and in aqueous dispersion polymerisation (Appendix, Type 5; Section 7.3.3) has shown that crosslinking of the core of the particle reduces the absorption of the incoming monomer so that polymerisation occurs in a layer on the outside of the particle.[100,101] This technique has important implications in the control of particle coalescence and other properties which lead to good film formation. A parallel technique has been used in conventional aqueous emulsion polymerisation by ICI[102] for the production of surface coatings.

11.3. Microgels

Microgels have been made for many years as impact modifiers for plastics by Du Pont, Rohm & Haas and ICI. The introduction of a difunctional monomer into aqueous emulsion polymerisation gives particles which, after precipitation, washing and drying, could be incorporated into plastics, or into solutions of polymers used in acrylic lacquers. The control over the interface and refractive index between the microgel and the matrix polymer was an area of major synthetic design effort to achieve impact modification and clarity in surface coatings.[103] PMMA films of similar tensile strength but greater plasticity may be made by the inclusion of low levels of microgels synthesised by the NAD technique.[104]

Microgels of condensation polymers have been made by incorporating multifunctional intermediates such as polyols into the preparation of polyester dispersions in aliphatic hydrocarbons.[106-108] These latex particles with functional groups such as —OH, —COOH or unsaturated residues have been used to make coating compositions which can be cured thermally by condensation with amino resins or by autoxidation with metal driers. Microgels made from other polymers and microgels of composite structure have also been described.[109-114]

Microgels made by the NAD process can be transferred to aqueous media where they are used for coatings applications in the automotive market.[115] This is achieved by growing a suitable copolymer layer onto the microgel which is then crosslinked using a suitable difunctional monomer. If this layer contains ionisable residues, such as carboxyl

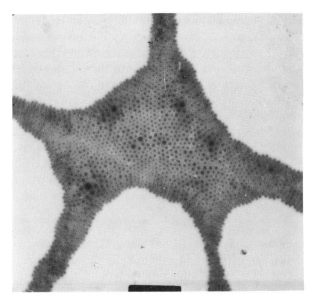

Fig. 4. Poly(styrene)/acrylic particles having 'core-shell' morphology (TEM, × 10 000, stained with phosphotungstic acid).

groups, in sufficient quantity the whole dispersion can easily be transferred into a dilute solution of aqueous alkali.

The method of aqueous dispersion polymerisation described in Section 7.3.3 can also be used to produce microgels and also microgels with encapsulating layers of polymer. By including difunctional monomers such as glycol dimethacrylate or allyl methacrylate, crosslinked polymer particles can be made with very high gel contents.[116] The growth of a secondary layer onto these crosslinked cores, such as PMMA or poly-2EHA, yields fine particles with core/shell morphology. An example of such a particle is shown in Fig. 4. In this case an aqueous dispersion polymerisation of butyl acrylate, styrene and allyl methacrylate was encapsulated by a second feed of methyl methacrylate to give a fine dispersion of heterogeneous particles.

Both types of microgel can be transferred from the aqueous medium to dispersion in other solvents such as toluene, methylisobutyl ketone, esters, etc., by azeotropic distillation.[117] Stability of the dispersions can be maintained either by the attached methoxypolyethylene glycol chains used for the aqueous dispersion or by the part dissolution of 'shell' polymer in the case of the two-phase, encapsulated particles.

11.4. Coloured Polymer Dispersions

Coloured polymer particles can be made by the inclusion of a coloured monomer into the dispersion polymerisation.[105] Similarly, stable non-polymerisable dyestuffs can be included in the monomer feed. In the case of aromatic polyester dispersions, colourants or dyestuffs can be included as solutions in the molten reactants prior to emulsification in a hydrocarbon and conversion to polymer.

Poly(ethylene terephthalate) particles of 0.2–$0.5 \mu m$ can be dyed in dispersion by a transfer dyeing process using Dispersol dyes.[118] The fine coloured polymer particle dispersions can be used as light-stable pigments in surface coatings. These methods of incorporating dyes can obviously be used to incorporate other chemicals such as pesticides, etc.

11.5. Dispersions of Polymers Containing Fillers and Pigments

The encapsulation of fillers in polymer by dispersion processes has been described by ICI.[119] In this process, polymer is precipitated onto the pigment in the presence of a dispersant and then monomer is grown in the polymer layer to give a substantial coating.

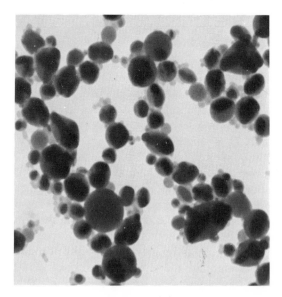

Fig. 5. TiO_2 encapsulated in nylon 11 (transmission EM, $\times$20 000).

Pigments may be milled into the molten intermediates for a condensation polymer, dispersed in a hydrocarbon and incorporated into the polymer as it is formed. Titanium dioxide pigment has been incorporated into poly(ethylene terephthalate) by this route.

A high level of mono encapsulation has also been achieved by the milling of TiO_2 pigment with 11-amino undecanoic acid in dispersion in an aliphatic hydrocarbon in the presence of a 'comb' graft copolymer in a process similar to the description in Section 11.4. The co-grind dispersion after conversion to polymer by the azeotropic removal of water gave polymer particles containing pigment particles (Fig. 5). Other fillers such as β-crystobalite, iron oxide, barium titanate, etc. can be similarly encapsulated, as shown in Table 4.

TABLE 4
Composite Particles made by Co-grind Process in Aliphatic Hydrocarbon

Polymer	Filler[a]	Proportions w/w	v/v	Total % wt dispersant	Particle size (μm)
Nylon 11	BC	1:1	3:1	14	0·2–1
Nylon 6·6	BC	1:1	3:1	3	2·25
Nylon 11	Alag	1:10	—	12	0·1–1
Nylon 6·6	CuI	5:1	1:1	0·75	2–8
Nylon 6·6	BaTit	5:1	1:1	0·65	2–5
Nylon 11	TiO_2	1:1	4:1	5	0·2–1
PET	TiO_2	4:1	16:1	2	20–50

[a] BC = β-crystobalite; Alag alumina cement (Lafarge); CuI = cuprous iodide; BaTit = Barium Titanate.

11.6. Continuous Addition Polymerisation in Hydrocarbons

A continuous process for the preparation of acrylic copolymers has been described by ICI.[120] The process is capable of giving 60% polymer dispersions or powders by using the heat of polymerisation to evaporate the aliphatic hydrocarbon.

Continuous processes for aqueous emulsion polymers are described in Chapter 2.

12. CONCLUSIONS

It is clear from the previous sections that fine polymer dispersions can be prepared in media of high or low dielectric constant such as water at one extreme and liquid inert gases at the other. The principles for the preparation of all types of polymer dispersion, from fluorinated hydrocarbons to inorganic polymers, are known and in most cases it is possible to make dispersions of submicron particle size. Both monodisperse and polymodal dispersions can be made in a controlled way. The main keys to these techniques lie in the understanding of the mechanisms of polymerisation coupled with the ability to tailor-make the appropriate dispersant which is firmly anchored to the particle and is capable of operating under the conditions of the polymerisation when large changes in polarity are taking place.

While this account is not intended to be a complete review of the literature whether academic or in patents, it is hoped that it will provide an overview of the wide range of methods which can be used to prepare polymer particles dispersed in liquids.

REFERENCES

1. Barrett, K. E. J. (ed.) *Dispersion polymerisation in organic media*, Wiley, London (1975).
2. Everett, D. H. and Stageman, J. F., *Colloid Polymer Sci.*, **255,** 293 (1977).
3. Everett, D. H. and Stageman, J. F., *Faraday Disc., Chem. Soc.*, **65,** 230 (1977).
4. Bamford, C. H., Barb, W. G., Jenkins, A. D. and Onion, P. F., *Kinetics of vinyl polymerisation by radical mechanisms*, Butterworths, London (1958).
5. Chatelain, J., *Brit. Polym. J.*, **5**(6), 457–65 (1973).
6. Hamielec, A. E. and Marten, F. L., *ACS symposium, polymer reactor processes, Series*, **104,** 43–9 (1978).
7. Albright, L. F., *Chemical Engineering*, **85,** July (1967).
8. Sanghvi, S. K., *Popular Plastics Annual*, 71 (1972).
9. Polnis, S. P. and Sharma, B. V., *Popular Plastics Annual*, 57 (1972).
10. Schildknecht, C. E., *High Polymer*, **29,** 88–105 (1977).
11. Kunstle, J. F. and Dehnke, M. K., *J. Radiation Curing*, **4**(2), 4–16, 18–32 (1977).
12. Matsumoto, M., Takakura, K. and Okaya, T., *High Polymer*, **29,** 198–227 (1977).
13. Burnett, G. M., *Kinetic mechanisms of polyreactions*, IUPAC Symposium Macromolecules Chem. Plenary Main Lecture (1971), pp. 403–15.
14. Okamura, S. and Urakawa, N., *Kogyo Kagaku Zasshi*, **53,** 303 (1950).
15. Henrice-Olivé, G. and Olivé, G., *Makromol. Chem.*, **68,** 219 (1963).
16. Young, L. J., *Polymer handbook*, 2nd edn, Wiley, London (1975), pp. 57–104.
17. Richards, D. H., *Polymer*, **19,** 109–11 (1978).

18. Athey, R. D., *Telechelic polymers—precursors to high solids coatings* Waterborne and Higher Solids Coatings Symposium, New Orleans (1981).
19. Thompson, M. W. and Waite, F. A., Brit. Pat. 1,096,912.
20. Tait, P. J., Dunn, D. J., Rooney, J. M., Young, R. N., Bevington, J. C., Bamford, C. H., Blackley, D. C. and Tideswell, B. M., *Specialist chem rep*; *macromolecular chemistry*, Vol. I, Royal Society of Chemistry, London (1980), pp. 3–80.
21. Herlinger, H., Hoerner, H. P., Druschke, F., Denneler, W. and Haiber, F., *Chemical reactions occurring during the solution, polycondensation of dicarboxylic acid chlorides with diamines in polar solvents*, Applied Polymer Symposium No. 21 (1973), pp. 201–14.
22. Barrett, K. E. J., Unpublished internal ICI Paints Division report (1968).
23. Kabanov, V. A., Papisov, I. M., Nedyalkova, T. S. I. and Avramchuk, N. K., *Polymer Science, USSR*, **A15**(9) 2259 (1973).
24. Shavit, N. and Cohen, J., *Polymerisation system in which the growing polymeric radical is absorbed on another macromolecule I 1–2*, Polymer Dept., Weizmann Institute of Science, Rehovot.
25. Ballard, D. G. H. and Bamford, C. H., *Proc. Roy. Soc. A*, **236**, 384 (1956).
26. Bamford, C. H. and Rice, R. C., *Trans. Faraday Soc.*, **61**, 2208 (1965).
27. Bamford, C. H., Block, H. and Imanishi, Y., *Biopolymers*, **4**, 1067 (1966).
28. Elias, H. G., *Polymerisation in organised systems*, Midland Macromolecular Monographs, Gordon and Breach, London, **3** (1977).
29. Bamford, C. H., *Developments in Polymerisation—2*, R. N. Hayward (ed.), Applied Science Publishers Ltd, London (1979).
30. Dubault, A., Casagrande, C. and Veyssié, M. *J. Physical Chemistry*, **79** (21) (1975).
31. Ackermann, R., Naegele, D. and Ringsdorf, H., IUPAC Symposia Madrid, 33–5 (1973).
32. Whitten, D. G. and Worsham, P. R., *Light induced polymerisation of surfactant styrene derivatives in monolayer films and assemblies*, Dept. of Chemistry, University of North Carolina.
33. Beredjick, N. and Birlant, W. J., *J. Poly. Sci. Part A-1*, **8**, 2807–18 (1970).
34. Dubault, A., Veyssié, M., Liebert, L. and Strezelecki, L., *Nature, Physical Science*, **245** (1973).
35. Day, D. R., Ringsdorf, H. and Lando, J. B., *ACS Polymer Preprints*, **19**(2) 176–8 (1978).
36. Naegele, D. and Ringsdorf, H., *J. Poly. Sci., Poly. Chem. Ed.*, **15**, 2821–34 (1977).
37. Loch, H. and Ringsdorf, H., *Makromol. Chem.*, **182**, 225–59 (1981).
38. Mupfer, B., Ringsdorf, H. and Schupp, H., *Makromol. Chem.*, **182**, 247–53 (1981).
39. Osmond, D. W. J. and Waite, F. A., Section 2.2 in Ref. 1.
40. Buscall, R. and Corner, T., *ACS symposium series, Emulsion polymers, emulsion polymerisation*, **165**, 157–169 (1981).

41. Corner, T., *Colloids and Surfaces*, **3**, 119 (1981).
42. Walbridge, D. J., Sections 3.7.3 and 3.7.4 in Ref 1.
43. Waite, F. A., *J. Oil Col. Chem. Assoc.*, **54**, 342 (1971).
44. Milkovitch, Corn Products International, USP 3,786,116; USP 3,832,423; USP 3,842,050; USP 3,842,057; USP 3,842,058; USP 3,842,059; USP 3,842,146; USP 3,846,393.
45. Dawkins, J. V. and Taylor, G., *Polymer*, **20**, 599 (1979).
46. Corner, T., *Advances in Polymer Science*, **62**, 95 (1984).
47. Osmond, D. W. J., Waite, F. A. and Walbridge, D. J., BP 117439.
48. Bromley, C. W. A. B. and Thompson, M. W., GB 2,039,497A.
49. Graetz, C. W., GB 2,051,096A.
50. Dawkins, J. V., *Polymer Colloids II*, R. M. Fitch (ed.) (1980), pp. 447–56.
51. Davies, S. P., Gibson, D., Parr, R. W. and Thompson, M. W., UK Patent Application 8,221,894.
52. Kendall and Co., USP 2,960,935 (1972).
53. Davies, S. P. and Thompson, M. W., UK Patent Applications 8,231,130; 8,303,585; 8,209,387; 8,213,880.
54. Goodall, A. R., Wilkinson, M. C. and Hearne, J., *J. Poly. Sci., Poly. Chem. Ed.*, **15**, 2193 (1977).
55. Fitch, R. M. and Kamath, Y., *Polymer Reprints*, **16**, 228–33 (1975).
56. Fitch, R. M. and Kamath, Y., *J. Colloid and Interface Science*, **54**, 6–12 (1976).
57. Barrett, K. E. J. and Thomas, H. R., *Kinetics and mechanisms of polyreactions*, IUPAC Int Symposium (1969), pp. 3375.
58. Dulux Australia Z/PV3190Z WP83 00151.
59. Pacific Vegetable Oil Co, USP 3,610,989.
60. Davies, S. P. and Thompson, M. W., UK Patent Application 8,214,675.
61. Davies, S. P. and Thompson, M. W., ICI Paints Division, unpublished results.
62. Doroszkowski, A., ICI Paints Division, private communication.
63. Hansen, F. K. and Ugelstad, J., *J. Poly. Sci., Poly. Chem. Ed.*, **16**, 1953–79 (1978).
64. Walbridge, D. J., Sections 3.5.1 and 3.5.2 in Ref 1.
65. Napper, D. H., *Trans. Faraday Soc.*, **64**, 1701 (1968).
66. Napper, D. H. and Netschey, A., *J. Colloid and Interface Sci.*, **37**, 528 (1971).
67. Dowbenko, R. and Hart, D. P., *Review Ind. Eng. Chem. Prod.*, **12** (1973).
68. Dowbenko, R., *Preparation and uses of non-aqueous dispersions*, 4th Int. Conf. in Organic Coatings Science and Technology, Athens (1978), pp. 129–46.
69. Dawkins, J. V., Taylor, G., Baker, P. and Collett, R. W., *ACS symposium series, emulsions and emulsion polymerisation*, **165**, 189–97.
70. Croucher, M. D., *Colloids and Surfaces*, **1**, 349–60 (1980).
71. ICI, BP 1,104,403.
72. Walbridge, D. J., Section 3.7.1 in Ref 1.
73. Thomas, H. R., Chapter 4, pp. 130–77 in Ref 1.
74. Rohm & Haas BP 956,453.
75. Rohm & Haas BP 1,329,062.

76. Firestone Tyre and Rubber Co, BP 1,007,476.
77. Firestone Tyre and Rubber Co, BP 1,008,180.
78. Firestone Tyre and Rubber Co, BP 1,008,188.
79. Hercules Powder Co, BP 1,165,840.
80. Hercules Powder Co, Belgium Pat 669,261.
81. Thompson, M.W., ICI Paints Division, unpublished results.
82. ICI Ltd, BP 1,412,584, South African Pat. 72/7635.
83. ICI Ltd, BP 1,095,931.
84. Murray, J. G. and Schwab, F. C., *Ind. Eng. Chem. Res. Dev.*, **21**, 93–6 (1982).
85. Barrett, K. E. J. and Thompson, M. W., Chapter 5, Section 5.5.1 in Ref 1.
86. Nicks, P. F. and Osborn, P., Ger Pat 2,215,732; BP 1,373,531.
87. Bentley, J. and Thompson, M. W., ICI Paints Division, unpublished results.
88. Union Carbide USP 4,000,218.
89. USM Corp, USP 3,917,741.
90. ICI BP 1,095,931; BP 1,095,932.
91. ICI B Pat Appln 17250; UK, 1,403,794.
92. ICI BP 1,399,344.
93. ICI BP 1,401,173.
94. Vincent, B., *J. Colloid and Interface Sci.*, **68**(1), 190 (1979).
95. Vrij, A., *J. Colloid and Interface Sci.*, **79**(1), 289 (1981).
96. Union Carbide USP, 3,632,669.
97. Firestone Tyre and Rubber Co, BP, 1,008,001.
98. ICI UK P 1,373,531.
99. Thompson, M. W., Chapter 5, pp. 229–31, in Ref. 1; BP 1,157,630.
100. ICI BP 1,364,698; Ger Pat 2,140,135.
101. West, E. J., ICI Paints Division, unpublished results.
102. ICI GB, 2,065,674.
103. ICI GB, 2,064,561A; 2,064,562.
104. ICI BP, 1,242,054.
105. BASF BP, 1,200,216.
106. ICI GB, 1,364,244; Ger Pat, 2,152,515.
107. ICI GB, 2,012,782.
108. ICI BP, 1,594,123.
109. ICI BP, 1,588,978.
110. ICI GB, 2,025,992.
111. ICI BP, 1,588,976.
112. ICI BP, 1,598,419.
113. ICI GB, 2,051,830.
114. Du Pont USP, 3,929,693.
115. ICI GB, 2,006,229A; GB, 2,073,609A; BP, 1,588,977.
116. ICI GB, 2,064,561A.
117. ICI GB, 2,064,562A.
118. ICI UK, 1,314,022.
119. ICI UK Pat, 1,453,713; UK, 1,506,222.
120. ICI BP, 1,234,395.
121. Cobbs, W. H. and Burton, R. L., *J. Poly. Sci.*, **10**, 275 (1972).

APPENDIX: METHODS USED TO

Type	Medium	Monomers	Locus of monomers	Locus of initiator or catalyst
1. Aqueous emulsion (ionic)	Water (acetone, alcohol)	Liquid vinyl and acrylic (addition polymerisation)	Emulsion droplet monomer migrates to micelle and growing polymer particle	In solution then macroradicals form part of micelles. Charged initiators, with stabilisation coupled to initiation
2. Aqueous emulsion (steric)	Water (acetone, alcohol)	Liquid vinyl and acrylic (addition polymerisation)	Emulsion droplet monomer migrates to micelle and growing polymer particle	In solution then macroradicals form part of micelles. Charged initiators, with stabilisation coupled to initiation (water-soluble, monomer-insoluble initiator)
3. Aqueous bead	Water	Liquid vinyl and acrylic (addition polymerisation)	Emulsion droplet	Emulsion droplet
4. Aqueous microbead	Water	Liquid vinyl and acrylic (addition polymerisation)	In fine emulsion droplet	Emulsion droplet
5. Aqueous dispersion I (precipitation and gel phase polymerisation)	Water/ alcohols/ acetone	Liquid or solid vinyl and acrylic (addition polymerisation)	In solution	In solution and then in particle as well
6. Dispersion II (precipitation and gel phase polymerisation)	Hydrocarbons and other organic liquid. Liquid gases	Liquid or solid vinyl or acrylic (addition polymerisation)	In solution	In solution and then in particle also. F. R. initiator anionic initiators, cationic initiators
7. Dispersion III (precipitation and aggregation)	Hydrocarbon and other inert organic liquids	Vinyl, acrylic epoxide lactams, isocyanates, *condensation* reaction monomers such as acid chlorides and diols	In solution	In solution
8. Dispersion IV (emulsified liquid monomer)	Hydrocarbon and other inert organic liquids	Liquid or easily meltable monomers giving *condensation* polymers, e.g. BHET	In droplet	In droplet

PREPARE POLYMER DISPERSIONS

Mechanism and locus of polymerisation	Dispersion stabilising agent	General comments
Initial solution polymerisation then in micelles and in growing polymer particle	Anions or cations with counter ion cloud (DLVO theory) + carbohydrates pluronics and tetronics as secondary colloids or stabilisers	$0.1–0.3\,\mu$m uniform particles SBR, PVA, PVC acrylates, etc. Particles have $-OH$, $-COOH$ and $-SO_3H$ $-NH_3–NR_3$ groups on surface from initiator action. Sensitive to ions, shear, pH
Initial solution polymerisation then in micelles and in growing polymer particle	Stabiliser precursor water soluble macromonomer.	$0.1–10\,\mu$m uniform or polymodal. Sterically stabilised ion-free stable to ions, shear pH, etc.
In droplet microbulk polymerisation	Polyacrylic acid 'granulating' agent at very low levels	$50–400\,\mu$m and aggregates. Method for preparing polyacrylate beads for moulding
In droplet microbulk polymerisation	Amphipathic 'comb' or double 'comb' graft polymers with polymerisable groups on backbone	$0.1–1\,\mu$m particles. Stable to ions, shear, pH
Initially in solution and then in precipitating polymer as well	Monofunctional water soluble oligomer, e.g. methoxy PEG-[meth]acrylate 'comb' graft polymers with water soluble side-chains	$0.01–10\,\mu$m particles. Acrylic latex with only steric or entropic stabilisation with no ionic double layer Stable to ions, shear, pH, freeze/thaw, etc.
In solution and then in particle also with pronounced autoacceleration and particles grow by monomer absorption	Oligomer or polymer with polymerisable groups or comb-graft molecules. Polydimethyl siloxane/styrene AB or ABA block copolymers.	'Dispersymer' (NAD) process $0.01–10$ μm uniform or polydisperse oleophilic particles
In solution with microparticles aggregating, no autoacceleration	Amphipathic graft polymers or precursors. Dispersant continually added or formed to cope with burying by aggregation	$0.1–10\,\mu$m particles with oleophilic surfaces. PVC, nylon 6, polyesters, polyformals
In droplet with elimination of byproducts such as water and ethylene glycol	Amphipathic graft copolymers of 'comb' or double 'comb' type optionally with groups reactable with polymer	$0.1–200\,\mu$m PET. dispersion. Stabiliser may be grafted to particle surface. Particles may be crosslinked during process if required. Linear crystallisable polymer dispersions need to be quenched to maintain stability

APPENDIX:—contd.

Type	Medium	Monomers	Locus of monomers	Locus of initiator or catalyst
9. Dispersion V (solid monomer)	Hydrocarbons and other inert organic liquids of suitable boiling point	Nylon 6:6 salt 11-amino undecanoic acid (condensation polymerisation)	Solid dispersed microparticle produced by attrition	In particle
10. Dispersion VI (solution of reactants emulsified)	Solutions of monomers in water, glycol, DMF DMSO NMP	Nylon salts, amino acids BHET acrylamide reactants for PES (condensation polymers)	In solution	In droplet
11. Dispersion VII (controlled precipitation of polymer)	Solvent, non-solvent for polymer	Acrylates, etc., condensation polymers	Monomer(s) (polymerised in solution)	In solution

Mechanism and locus of polymerisation	Dispersion stabilising agent	General comments
In particle which goes through a semi-liquid state with byproducts easily removed by azeotropic distillation owing to fine particle size not acting as barrier to rapid diffusion	Comb-graft amphipathic polymer with reactive groups	0·1–100 μm monomer, nylon 6.6, 11, etc. ground to required fineness and particle size dependent on this. Dispersion quenched to maintain stability in potentially crystallisable polymers. Process suitable for encapsulation and random/random copolymers.
In droplet as solvent is removed by azeotropic distillation	Amphipathic 'comb' and 'double comb' graft copolymers with polymerisable groups	0·5–100 μm suitable for making copolymers
Solution-polymerisation then polymer precipitated by adding solution to non-solvent containing dispersant	Preformed amphipathic graft copolymers	Relatively dilute dispersions without solvent stripping

Chapter 2

Mechanisms and Kinetics of Emulsion Polymerisation

GARY W. POEHLEIN

School of Chemical Engineering, Georgia Institute of Technology, Atlanta, Georgia, USA

1. INTRODUCTION

Emulsion polymerisation is a reaction technique which is used to produce polymer colloids from unsaturated monomers via free radical reactions. The emulsion polymerisation system is distinguished from bulk, solution and suspension polymerisations by the high degree of subdivision of the colloidal polymerisation sites. Individual or small numbers of free radicals can be isolated in these small monomer-swollen polymer particles and, in most cases, high molecular weight polymer can be produced at very high rates.

The relatively low viscosity of the reaction media facilitates rapid heat transfer and easy product handling. The latex form of the product is directly useful, as coatings, adhesives and additives for other products. Some of the most important applications such as synthetic rubbers and moulding plastics, however, require isolation of the polymer. The recipe ingredients which are necessary to carry out the reaction and to stabilise the colloidal particles can cause application problems. This is one of the potential disadvantages of emulsion polymerisation. Nevertheless, the range of polymer latex applications has expanded considerably in the past 10–20 years and production volumes have increased dramatically.

Research and development activities have also increased significantly since the mid 1960s. Advances have been made in understanding preparation techniques, chemical and physical properties, characterisa-

45

tion and application performance. The purpose of this chapter will be restricted to the preparation area. The three topics to be reviewed include: (1) chemical and physical reaction mechanisms; (2) polymerisation kinetics; and (3) reactor design and operation. An in-depth treatment of all these topics will not be possible in a single chapter so the presentation will be designed to review most of the important concepts in a qualitative manner. An extensive literature is available for those who wish to pursue emulsion polymerisation in more detail.

2. FREE RADICAL REACTIONS AND KINETICS

The important chemical reactions in free radical polymerisation and the rate expressions for these reaction steps are shown below.

2.1. Initiation

$$I_2 \xrightarrow{k_d} 2I \cdot \qquad R_d = k_d[I_2]$$

$$I \cdot + M \xrightarrow{k_{pI}} R_1^\cdot \qquad R_i = k_{pI}[I \cdot][M]$$

The rate-controlling step is usually initiator decomposition and the rate of initiation, R_i, is normally expressed as shown by eqn (1).

$$R_i = 2fk_d[I_2] \qquad (1)$$

where f is an initiator efficiency factor and k_d is the decomposition rate constant. The reactions shown here would apply to most thermal initiators such as the organic peroxides and the persulphate ion. Redox initiation systems involve several chemical species and the rate expressions can be more complex.

2.2. Propagation

$$R_r^\cdot + M \xrightarrow{k_p} R_{r+1}^\cdot \qquad R_p = k_p[M][R \cdot] \qquad (2)$$

where $[R \cdot] = \sum_{r=1}^{\infty} [R_r^\cdot]$ is the total concentration of free radicals. The assumption inherent in the above rate expression is that the propagation rate constant, k_p, is not a function of the radical length, r. Also R_p is usually assumed to be the polymerisation rate since the relative amount of monomer consumed in the initiation step is small. A second way to justify this equality is to assume $k_{pI} = k_p$, $f = 1 \cdot 0$ and include $[I \cdot]$ in the $[R_r^\cdot]$ summation.

The propagation rate expressions become more complex when more than one monomer is used because several species of free radical and monomer exist, and different rate constants are used for each of the possible propagation reactions. Another possible complication is introduced when radical transfer reactions occur. In these cases one needs to make an assumption about the reactivity of the newly formed free radical.

2.3. Termination

$$\text{R}_r^{\cdot} + \text{R}_s^{\cdot} \xrightarrow{k_{tc}} \text{P}_{r+s} \qquad R_{tc} = k_{tc}[\text{R}\cdot]^2$$
(coupling)

$$\text{R}_r^{\cdot} + \text{R}_s^{\cdot} \xrightarrow{k_{td}} \text{P}_r + \text{P}_s \qquad R_{td} = k_{td}[\text{R}\cdot]^2$$
(disproportionation)

The total rate of termination, which is the sum of the reactions shown above, is given by eqn (3).

$$R_t = k_t[\text{R}\cdot]^2 \qquad (3)$$

where $k_t = k_{tc} + k_{td}$. Although the reaction rates can be added, the products are different because one of the polymer molecules formed in the disproportionation reaction contains a terminal double bond and can, therefore, participate in later propagation reactions to form a branched molecule.

One of the important problems which must be handled when using eqn (3) in reactor model development is the fact that k_t can be strongly dependent on the mobility of the reacting radicals and thus on the conversion of the reaction as well as on temperature. This phenomena is called the 'gel' or 'Tromsdorff' effect. At high conversion the propagation reaction can also become diffusion-controlled. Hamielec and MacGregor[1] and Sundberg et al.[2] have presented semi-empirical models for the influence of polymer molecular weight and free volume on the propagation and termination constants.

2.4. Transfer Reactions

$$\text{R}_r^{\cdot} + \text{TX} \xrightarrow{k_{tr}} \text{P}_r + \text{T}\cdot \qquad R_{tr} = k_{tr}[\text{R}\cdot][\text{TX}]$$

There is no net change in the number of free radicals in a transfer reaction. Thus if the reactivity of the new radical, T·, is the same as $\text{R}_r^{\cdot}$, the transfer reaction will not alter the polymerisation rate, except perhaps in emulsion polymerisation if T· diffuses from the reaction site into the continuous phase.

Transfer reactions can occur with a wide variety of chemical species including monomer, emulsifier, solvent, polymer and chain transfer agents or modifiers. Chain transfer agents such as mercaptans are added to some emulsion polymerisation systems (e.g. SBR) to control molecular weight and to reduce branching and crosslinking. The reactivity of the radical formed in the transfer reaction can reduce the rate in two ways. If the radical is very unreactive, the transfer agent is called an inhibitor or retarder and the reaction is effectively a first-order termination. This phenomenon can occur in bulk, solution, suspension or emulsion polymerisation. If the radical formed is highly reactive, the rate should not be influenced except in emulsion polymerisation. A small mobile free radical may diffuse out of a colloidal latex particle into the aqueous phase where the monomer concentration is small. This transport phenomenon will cause a reduction in rate. Radical transport phenomena will be considered in more detail in Section 3.3.

3. POLYMERISATION RATE RELATIONSHIPS

The rate of polymerisation, as expressed by the propagation rate, is dependent on the first power of the free radical concentration $[R·]$. This concentration can be eliminated from the rate expression by equating the rates of initiation and termination.

$$R_i = R_t = k_t[R·]^2 \quad \text{or} \quad [R·] = (R_i/k_t)^{\frac{1}{2}} \tag{4}$$

Equation (4) results by assuming that $[R·]$ does not vary significantly with time and that no inhibiting reactions (first-order termination) are present. Substitution of eqn (4) into the R_p relationship yields the rate expression which is normally used for bulk, solution and suspension polymerisations.

$$R_p = k_p[M][R·] = k_p[M](R_i/k_t)^{\frac{1}{2}} \tag{5}$$

The units of the various parameters are based on a unit volume of reaction mixture: R_p (moles monomer polymerised/time–volume), $[M]$ (moles monomer/volume), $[R·]$ (moles radicals/volume) and k_p (volume/mole–time).

A number of different considerations are present in the rate expression for emulsion polymerisation which is given by eqn (6).

$$R_p = k_p[M_p]\left(\frac{\langle \bar{n} \rangle N}{N_A}\right) \tag{6}$$

First, the rate of polymerisation is usually based on a unit volume of continuous phase (moles monomer converted/time–volume of water). Second, the monomer concentration, $[M_p]$, is the concentration at the reaction site within the polymer particles. Lastly, the free radical concentration $[R\cdot]$ is expressed by $\{(\langle \bar{n} \rangle N)/N_A\}$, where N is the particle concentration (number of particles per volume of aqueous phase), N_A is Avogadro's number and $\langle \bar{n} \rangle$ is a double average. $\bar{n}$ is the average number of free radicals per particle among particles of a specific size and the brackets, $\langle \rangle$, reflect an average over the particle size distribution.

The propagation rate constant can be obtained from the literature or from bulk or solution polymerisation experiments. Thus the prediction of emulsion polymerisation rates involves obtaining numerical values for $[M_p]$, $\langle \bar{n} \rangle$ and N. Each of these parameters will be discussed separately in the remainder of this section.

3.1. Monomer Concentration, $[M_p]$

Several assumptions are inherent in the use of the term $[M_p]$ in eqn (6). First, the monomer-swollen polymer particles are assumed to be the dominant locus of polymerisation. Ugelstad et al.[3] clearly demonstrated that significant polymerisation can occur in monomer droplets if special emulsification systems are employed. Durbin et al.[4] showed that some polymerisation occurs in monomer droplets even with classical styrene recipes. Equation (6) is not a valid rate expression if significant polymerisation takes place in the monomer droplets or in the continuous phase.

When monomer droplets are present the value of $[M_p]$ is normally computed based on an assumption of equilibrium swelling of the polymer particles. The free energy of mixing is the driving force for swelling when a particle is exposed to a system containing free monomer. The resisting force is the energy required to expand the surface area of the particle. Morton et al.[5] developed a quantitative model for the prediction of equilibrium swelling by using the Flory–Huggins equation ($\Delta \bar{F}_m$—mixing) and the Gibbs–Thomson equation

($\Delta \bar{F}_i$—interfacial) for the two free-energy terms ($\Delta \bar{F}_m + \Delta \bar{F}_i = 0$). Morton *et al.* showed that, in the normal latex size range, the swelling ratio was not a strong function of particle size. Hence most workers have used a constant value of $[M_p]$ in rate expressions and kinetic studies. The equilibrium monomer concentration does, however, vary slightly with particle size, even for rather standard latices. Tseng *et al.*[6] re-examined this problem for the poly(styrene)–styrene and poly–MMA–MMA systems. In order to account for the particle size effect, eqn (6) should be rewritten as follows:

$$R_p = k_p \left(\frac{\langle [M_p]\bar{n} \rangle N}{N_A} \right) \qquad (7)$$

where the product $[M_p]\bar{n}$ is averaged over the size distribution. This is not normally done and $[M_p]$ values measured at equilibrium in non-reacting systems are used in rate expressions.

When the free-monomer phase disappears the amount of remaining monomer can be determined by mass balance. This remaining monomer is also assumed to exist at equal concentrations within particles of different size.

In comparison to $\langle \bar{n} \rangle$ and N, the quantitative determination of $[M_p]$ for most emulsion polymerisation systems will be relatively simple. Experimental equilibrium values have been published and measurement techniques are not difficult. In addition, theoretical models are available for obtaining predicted values. The paper by Tseng *et al.*[6] provides a more detailed review of this problem area.

Recent work, with recipes which contain minor ingredients to promote swelling, has achieved very high monomer : polymer swelling ratios and very large latex particles. This work has been reviewed by Ugelstad *et al.*[7,8] and the kinetic treatment described above will not be valid without consideration of other thermodynamic and perhaps mass transport effects.

3.2. Particle Concentration, N

Prediction of the number of particles that will be formed as a function of the recipe used and reactor operation policy is the most difficult problem in emulsion polymerisation kinetics. Theoretical models are available and, in some cases, reasonable predictions are possible. Most of these models, however, contain unknown adjustable parameters and many have not been tested significantly with carefully designed experiments. Hence, experimental determination of N will be necessary for most recipes if accurate values are needed. Theoretical work, some of which is reviewed below, can provide good predictions for some simple systems and may be helpful for model extrapolations in more complex systems.

Smith and Ewart[9] derived the relationship given by eqn (8) for particle concentration in an S–E Case 2 batch reaction.

$$N = kR_i^{0.4}[S]^{0.6} \tag{8}$$

where [S] is the emulsifier concentration and k is a constant. The physical mechanism considered for particle nucleation by Smith and Ewart involved the transport of water-borne free radicals into monomer-swollen emulsifier micelles. These newly formed particles would then grow by polymerisation, exposing more organic surface for emulsifier adsorption. Smith and Ewart postulated that the micelles which are not 'stung' by free radicals would break up rapidly to supply the emulsifier for the expanding particle surfaces. Particle nucleation would stop after the micelles had all broken up.

These same concepts were used by other workers[10-12] to predict the number of particles that would be formed in a single steady-state continuous stirred-tank reactor (CSTR). The CSTR particle model is given by eqn (9).

$$N = k_c[S]^{1.0}\theta^{-0.67} \tag{9}$$

where k_c is a constant and θ is the mean residence time (reactor operating volume divided by the volumetric throughput rate) of the CSTR. The differences between eqns (8) and (9) result because of differences between CSTRs and batch reactors. The chemical and physical mechanisms considered in the development of the equations are identical.

Roe[13] questioned the role of micelles in the nucleation process and demonstrated that eqn (8) could be derived if the micelles were simply considered to be reservoirs which supply emulsifier for the expanding particle surfaces. He suggested aqueous phase nucleation, such as reported earlier by Priest,[14] as a feasible particle formation process.

Brooks[15] examined particle formation in CSTRs in more detail. Among other things, he considered the possibility that micelle break-up was not very rapid and demonstrated that 'satisfactory predictions for the particle number can be obtained without assuming that the polymer particles are always saturated with surfactant'. Brooks also reported briefly on particle number oscillations.

The simple particle number models presented above have been tested almost exclusively with styrene emulsion polymerisation experiments. Fitch and co-workers[16-18] formulated a theory for nucleation based on the assumption that particles are formed in the continuous phase by precipitation of oligomeric radicals which have grown beyond a critical chain length. The Fitch model for homogeneous nucleation, in general form, is given by eqn (10).

$$\frac{dN}{dt} = R_i - R_c - R_f \tag{10}$$

where R_i is the rate of generation of free radicals, R_c is the rate of capture of these free radicals by particles and R_f is the rate of particle flocculation.

The problem of the prediction of N becomes that of integrating eqn (10) over the course of the reaction. Fitch and co-workers treated this problem for various cases, and Hansen and Ugelstad[19] developed more extensive models and solutions. Nevertheless, a great deal remains to be done to permit the quantitative prediction of N from a knowledge of recipe and reactor variables. The area of nucleation in monomer droplets has also been reviewed by Hansen and Ugelstad.[20] They examined nucleation in droplets in competition with the micellar and homogeneous mechanisms.

3.3. Free Radical Concentrations in the Particles, $\langle \bar{n} \rangle$

The problem of predicting the average number of free radicals per particle requires, at least in some cases, that the particle concentration, N, be known. Thus in this section it will be assumed that N is a parameter whose numerical value has been computed or measured. The theories presented here are valid when a water-soluble initiator is used with a fixed particle population, i.e. after the nucleation period.

A balance equation for the number of particles containing n free radicals is given by eqn (11).

$$\frac{dN_n}{dt} = \frac{\rho_A}{N}[N_{n-1} - N_n] + k_d[(n+1)N_{n+1} - nN_n]$$

$$+ \left(\frac{k_t}{v}\right)[(n+2)(n+1)N_{n+2} - n(n-1)N_n] \tag{11}$$

where ρ_A is the rate of absorption of free radicals by the particles, $N = \sum_{n=0}^{\infty} N_n$, k_d is a coefficient for transport of free radicals out of the particles, k_t is the rate constant for bimolecular termination within the particles, and v is the volume of the monomer-swollen particles. Various workers, starting with Smith and Ewart,[9] have obtained limited solutions for the system represented by eqn (11). These solutions are usually based on a steady-state analysis $(dN_n/dt = 0)$ of monodisperse latices. The solution presented by O'Toole[21] is given by eqn (12).

$$\bar{n} = \left(\frac{a}{4}\right) \frac{I_m(a)}{I_{m-1}(a)} \tag{12}$$

where $\bar{n}$ is the average number of free radicals among the particle population, and m and a are dimensionless parameters defined as:

$$m = k_d/vk_t \tag{13}$$

$$a = \sqrt{(8\alpha)} \qquad \alpha = \rho_A v/Nk_t \tag{14}$$

O'Toole also presented relationships for the distribution of n values among the particles.

Utilisation of eqn (12) is not straightforward because ρ_A, the rate of absorption of free radicals, cannot be calculated independently. It is influenced by the rate of initiation, by termination in the aqueous phase and by reabsorption of radicals that were previously transported out of the particles. Equation (15), originally given by Ugelstad et al.,[22] gives the relationship between these phenomena.

$$\alpha = \alpha' + mn - Y\alpha^2 \tag{15}$$

where $\alpha' = \rho_i v/Nk_t$, ρ_i is the rate of initiation (a parameter that can be determined independently), $Y = 2Nk_tk_{tw}/k_a^2v$, k_{tw} is the termination coefficient in the water phase and k_a is a radical absorption coefficient.

Ugelstad et al.[22] used numerical techniques with a digital computer to obtain solutions to eqns (12) and (15). Typical results are shown in Fig. 1. If a', m and Y are known, one can obtain a value of $\bar{n}$ from Fig. 1. The three special cases considered by Smith and Ewart[9] are embodied in Fig. 1. The straight horizontal line with $m = 0$ and log $\alpha' < -1$ is analogous to S–E Case 2 ($\bar{n} = 0.5$). The lower left side with higher values of m includes Case 1 ($\bar{n} \ll 0.5$) and the upper right side represents Case 3 ($\bar{n} \gg 0.5$).

Values of $\bar{n}$ obtained from Ugelstad plots reflect a fixed set of conditions for a monodisperse latex. During the course of a batch reaction the location of the point ($\bar{n}, \alpha', m, Y$) will prescribe a path in the solution space. In order to describe the performance of a batch or

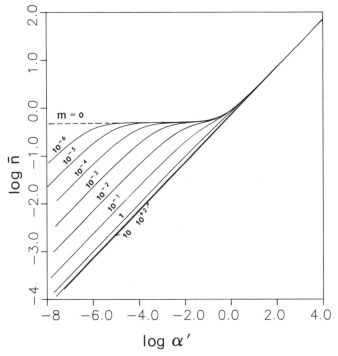

Fig. 1. Ugelstad plot for $Y = 0$.

semi-batch reactor one needs to determine this path in a quantitative manner.

If the latex is not monodisperse the influence of particle size on free radical transport and internal reactions must be considered. Poehlein *et al.*[23] examined this problem for a CSTR which is fed with a monodisperse seed latex. The latex effluent from such a reactor will have a broad particle size distribution. The transport of free radicals was assumed to follow the predictions of classical diffusion theory. Typical results are shown in Figs 2, 3 and 4.

Figure 2 is analogous to the single curve for $m = 0$ in Ugelstad's plot. The parameters γ, β, α'_c and Y_c are given by:

$$\gamma = \left(m \frac{v_s}{v}\right)^{\frac{1}{3}} \qquad \beta = v_s/\theta K_1[M_p] \qquad (16)$$

$$\alpha'_c = \rho_i v_s/Nk_t \qquad Y_c = \frac{2Nk_t k_{tw}}{[4\pi D_w N(\frac{3}{4}\pi)^{\frac{1}{3}} v_s^{\frac{1}{3}}]^2 v_s} \qquad (17)$$

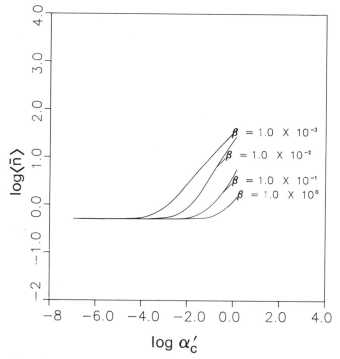

Fig. 2. Influence of β and α_c on the average number of free radicals per particle in a CSTR for $\gamma = 0$ and $Y_c = 0$.

where v_s is the volume of the monomer-swollen seed particles and K_1 is a particle growth parameter and D_w is the radical diffusion coefficient in the water phase. Figure 2 illustrates that smaller values of β (longer mean residence times, θ) cause earlier deviations from S–E Case 2 kinetics.

Figure 3 is analogous to Fig. 1 for a fixed value of β. The influence of the radical desorption parameter, γ, is the same as that of m. Figure 4 shows the influence of radical transport from particles on the particle size distribution of the effluent latex. In the absence of radical desorption, a single-peaked distribution is obtained whereas a double-peaked distribution is predicted for some values of γ. Please note that the dimensionless diameter, d/d_s, is used for this graph.

Gilbert and co-workers[24,25] used matrix methods to obtain several restricted solutions to the system of equations represented by eqn (11) without making the steady-state assumption. Among other conclusions, they suggest that the free radical capture efficiency of the latex

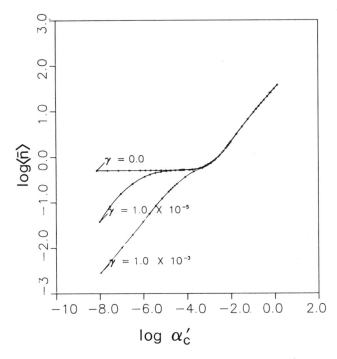

Fig. 3. Influence of γ and α_c on $\langle \bar{n} \rangle$. $\beta = 1 \cdot 8 \times 10^{-3}$, $Y_c = 0$.

particles can be quite low due to termination in the aqueous phase. This factor is considered, through the parameter Y, in the Ugelstad treatment but Y has usually been set equal to zero for data analysis.

To summarise, $\bar{n}$ or $\langle \bar{n} \rangle$ can be computed by several theoretical models if appropriate transport or kinetic parameters are known. Since this is not always the case, these theories can be used, with experimental data, for parameter estimation. The theories can also be applied for interpolation or extrapolation of experimental results.

4. REACTOR CONSIDERATIONS

Emulsion polymerisation reactions are carried out in batch, semi-batch and continuous reactors. The type of reactor system and the way it is

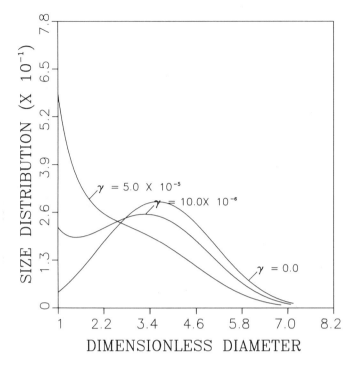

Fig. 4. Influence of radical desorption on the particle size distribution in the latex product of a CSTR. $\beta = 7 \cdot 1 \times 10^{-3}$, $\alpha'_c = 3 \cdot 5 \times 10^{-6}$, $Y_c = 0$.

operated can have very significant effects on the polymerisation kinetics and the characteristics of the latex produced. The purpose of the remainder of this chapter is to review the major differences between reactor systems and to discuss how these differences can influence the kinetics and product characteristics.

Reactors used for most polymerisation systems tend to be similar in that they are stirred vessels. Some applications have been suggested for tubular reactors, both in conjunction with CSTRs and in recirculation-loop configurations. The different operational modes are:

Batch: all recipe ingredients are added at the beginning of a batch reaction and the product is removed after the desired conversion is obtained.

Semi-batch: some components of the recipe are withheld from the
initial charge and added later, in a programmed manner, in semi-batch operation. Later additions often
involve part of the monomer charge, but sometimes
include other ingredients such as water, emulsifier and
initiator components.

Continuous: reagent and product streams are added and removed at
constant rates from continuous systems. The recipe
ingredients are normally all added at a single location
but this need not be the case in multiple-reactor continuous processes.

The differences in performance of the various reactor processes are
best illustrated by considering the influence of process type on important kinetic and latex parameters. Kinetic parameters include polymerisation rate, inhibitor effects and particle nucleation. Latex characteristics include particle size distribution, copolymer composition, particle
morphology, polymer molecular weight and the amount of coagulum
formed. Each of these important areas is reviewed briefly in the
remainder of this chapter.

4.1. Polymerisation Rate

The rate of polymerisation in a batch reactor is established by the
recipe used and the reaction temperature control. In the absence of a
seed latex the reaction passes through at least three intervals: (1)
particle nucleation; (2) particle growth in the presence of monomer
droplets; and (3) conversion of monomer to polymer in the swollen
particles after the monomer droplets have disappeared. A second
particle nucleation phase has sometimes been detected, presumably
because emulsifier becomes available when the particles shrink during
Interval 3 and because initiator end-groups add to the stabilising
ability of the system.

The course of a batch reaction can be controlled by three methods:
recipe formulation, heat removal and reaction termination. Such reactions tend to proceed with very non-uniform rates and heat generation.
Batch-to-batch variations can be significant.

One motivation for employing semi-batch operation is to obtain
better control of polymerisation rate. The monomer-addition technique can achieve very good rate and thus heat-release control. In such
systems the polymerisation can take place in monomer-starved parti-

cles and polymer transfer reactions leading to branching and/or cross-linking become more important.

The rate of polymerisation in a steady-state continuous process is constant but may vary among different reactors in a multiple-reactor system. This constant rate and heat-load is often cited as an advantage of continuous reactors.

4.2. Inhibitor Effects

Inhibitors are present in most monomers to prevent premature polymerisation during processing, shipping and storage. In addition, trace chemicals in other recipe ingredients, e.g. oxygen in water, can also inhibit or retard polymerisation.

When inhibitors are present in a batch reactor they delay the start of polymerisation but the reactions proceed normally after the inhibitor is spent. Such is not the case with semi-batch or continuous reactors because inhibitors can be present in the feed streams which are added to a reacting mixture.

If an inhibitor is present in semi-batch recipe ingredients the polymerisation in the initially charged ingredients will be delayed as is observed for batch reactors. When the flow of the streams containing the remainder of the recipe begins, however, any inhibitors present will serve to reduce the rate of initiation within the reactor. In extreme cases the polymerisation may be completely stopped. Likewise when the programmed additions are finished the initiation rate may increase rapidly. This phenomenon can cause excessive heat generation and a loss of temperature control. In extreme cases the continued addition of a dilute inhibitor stream may be required.

Continuous reactor systems often are comprised of a number of CSTRs (sometimes as many as 12 to 15) connected in series. Operation procedures usually include pumping all the recipe ingredients into the first reactor and removing the latex product from the final reactor. In such cases the inhibitor entering the first reactor with the reagent streams will reduce the rate of initiation in that reactor. If inhibitor concentrations are sufficiently high polymerisation reactions will not be initiated in the first reactor. A second, and perhaps less dramatic, possibility is that the initiation rate will be higher in the second CSTR in the series. This can come about because significant initiator is carried over into the second reactor whereas essentially all of the free radicals generated in the first reactor react with inhibitor. A Bayer patent[26] for chloroprene emulsion polymerisation discusses this problem and indicates that an additional inhibitor stream is sometimes required to achieve adequate temperature control in the second reactor.

4.3. Particle Nucleation

The total number of particles formed and the time of their formation can vary greatly among reactor types, even when the same overall recipe is employed. The particle formation period (Interval 1) in a batch reactor is often completed at very low monomer conversion levels (3–10%). Smith and Ewart's[9] simple particle formation continues until sufficient surface area is generated for adsorption of all of the added emulsifier. Since the particles in a batch reactor are small at the end of this nucleation period, a relatively large number can be stabilised. As these particles grow, more surface area will be generated and the system will not normally remain saturated with emulsifier.

Particle nucleation can vary considerably with operation procedures in semi-batch reactor systems. If the initial charge contains all ingredients except part of the monomer, particle nucleation will be very similar to that in a batch reactor. If, however, some of the emulsifier is withheld for later programmed addition, fewer particles will be formed. If part of the initiator is withheld, the lower rate of free radical generation during the nucleation period can also result in fewer particles.

Particle generation in continuous systems comprised of CSTRs is almost always confined to the first reactor in the series. Nucleation in a steady-state CSTR must occur in the presence of a large number of particles. Although Brooks[15] has shown that particle nucleation can take place in CSTRs which are not saturated with emulsifier, the average amount of emulsifier per particle is higher in the CSTR effluent than in a batch reactor at the end of Interval 1. This is simply because the average particle size is larger in these two comparative points in the reaction. Thus if the same recipe is employed one should expect to form fewer particles in a CSTR system. Theoretical predictions, based on S–E Case 2 kinetics, suggest that a single CSTR, even when operated in a manner to produce a maximum number of particles, will only generate about 60% of the particle concentration that would result from the same recipe in a batch reactor.

The mean residence time, θ, within the CSTR can have a significant influence on the number of particles formed. There is no analogous parameter in batch reactors. If the mean residence time is large, average particle size in the reactor is also large and the emulsifier added will stabilise a small number of particles. If the mean residence time is small, particle nucleation will be limited by free radical initiation. Thus some value of θ will provide a maximum number of particles. This is the maximum mentioned above.

Several techniques are available to increase the particle concentration in continuous systems. First, the recipe can be changed to include more emulsifier. Second, a continuous tubular reactor can be placed in front of the first CSTR to generate the particles. Particle nucleation in the tube can be nearly the same as in a batch reactor if the tube residence time is large enough to finish the Interval 1 part of the polymerisation. A third, and nearly equivalent method, involves the use of a seed latex which has been prepared in a separate batch or semi-batch reactor.

A last factor to be considered when comparing particle nucleation in various types of reactor is the influence of recipe and reactor variables. The earlier discussion on prediction of the particle concentration, N, clearly indicates that the influence of initiation rates and emulsifier concentration may be quite different, even with identical recipes. This observed phenomenon is one reason why those interested in commercial production of latex products should conduct product and process development studies in small-scale reactor systems of similar design.

4.4. Particle Size Distribution

Latex particle size distribution (PSD) can be an important parameter for application performance. PSD in a latex product will depend strongly on the reactor system. The latex product from a batch reactor is usually of rather narrow PSD because all the particles are formed early in the reaction. These particles have nearly the same age and hence the same size. In some cases nucleation later in the batch cycle can generate broader PSDs.

Operation policies can influence the PSD in a semi-batch system. If the particle nucleation period is short and confined to the initial part of the reaction, the PSD can be very narrow. If, however, the nucleation period is expanded or divided by the programmed addition of stabiliser, broad PSDs can be produced. Such techniques are sometimes employed to manufacture high-solids latex with controlled rheology.

The PSD of the latex produced in a CSTR system will also certainly be broader. The distribution of residence times in a single CSTR is given by eqn (18).

$$f_1(t) = \frac{1}{\theta} \exp(-t/\theta) \tag{18}$$

where $f_1(t)$ is the residence time density function and t is time. The particle age distribution in the effluent will be the same broad function. This will, of course, lead to a broad PSD. If N equal-size CSTRs are connected in series, the residence time density function is:

$$f_N(\tau) = \frac{N(N\tau)^{N-1}}{(N-1)!} \exp(-N) \tag{19}$$

where $\tau = t/N\theta$. The distribution of particle ages, as expressed by eqn (19), becomes narrower as N increases. Hence, the styrene–butadiene rubber product from systems of 12 to 15 reactors can be reasonably narrow. Narrow distribution latex can also be produced in once-through continuous tubular reactors. Tubular reactors which involve circulation of the reacting latex in a loop should produce a product with a broad PSD.

4.5. Copolymer Composition
When a copolymerisation reaction is carried out in a batch reactor the composition of the copolymer formed changes with time because of differences in monomer reactivities. The first copolymer formed is rich in the more reactive monomer and the last product contains a higher proportion of the least reactive monomer. The water solubility of the monomer can also be a factor in this phenomenon because of the initiation and oligomer reactions taking place in the water phase.

Semi-batch reactors, with programmed addition of the more reactive monomer, have been used to change the normal course of batch copolymerisation reactions. Monomer addition may be programmed to minimise composition drift or to produce considerable drift and non-uniform particles. Bassett and co-workers[27,28] have recently reported on a simple technique for producing particles with controlled non-uniformity.

The composition of copolymer produced in a steady-state CSTR will not, except for small stochastic variations, change with time. However, the copolymer produced in different reactors of a CSTR train will usually be different. The polymer formed in the first reactor will contain a higher fraction of the more reactive monomer than that formed in downstream reactors. Copolymer composition can be controlled by feeding monomer at various points along the reactor system. The relative flow rates of these intermediate feed streams can be controlled to achieve a variety of composition profiles within the reactor train.

4.6. Particle Morphology

Particle morphology of copolymer latices can be an important parameter in establishing application performance. As mentioned in Section 4.5, reactor type and operation procedures can have a significant influence on copolymer composition and, hence, on particle morphology. When a uniform composition or morphology is desired, programmed monomer feeding of a semi-batch reactor or intermediate feed streams in a CSTR system can be used.

If non-uniform particles are desired, programmed feeding in a semi-batch system can be used if a simple batch reaction will not yield the desired morphology. Bassett and Hoy[27] describe a simple process for achieving non-uniform particles. When such particles are produced in a semi-batch system a variety of results can be expected. If all particles are nucleated early in the reaction, the development of morphology will depend on monomer addition procedures and all particles should be very similar. If, however, a second period of particle nucleation occurs later, the new particles will have different compositions and morphologies.

The products from a CSTR system can also be quite varied, depending on operating procedures and intermediate feed streams. The distribution of residence times for each individual reactor is quite broad, as indicated earlier by eqn (18). This fact will be important in establishing differences in morphology among the particles in the final effluent stream. This issue can be illustrated by considering a two-reactor CSTR series in which the copolymer being formed in the first reactor is rich in monomer 1 and that in the second reactor is rich in monomer 2.

The partially converted latex from the first reactor would be comprised of particles with broad size and age distributions. If no new particles are formed in the second reactor, which would be the normal situation, the particles leaving reactor 1 would be overcoated with copolymer rich in monomer 2 in reactor 2. The latex effluent from the second reactor would have a narrower size distribution, but particles of the same size could have quite different structures. A particle with a small residence time in reactor 1 would have a small core and a large shell. On the other hand, a particle that was large when it was removed from reactor 1 may only accumulate a thin shell if its residence time in reactor 2 was small.

Such differences between particles of the same size would tend to become smaller as the number of reactors in the CSTR system increases. Nevertheless this effect of the individual reactor age distributions would always be present to some degree.

4.7. Polymer Molecular Weight Characteristics

The number-average molecular weight ($\bar{M}_n$) of the product of a polymerisation reaction depends on a number of factors including the relative rates of the propagation, termination and transfer reactions, and the functionality of the monomer. The value of $\bar{M}_n$ increases when the propagation rate increases relative to termination or transfer rates with small molecules. Transfer reactions with polymer molecules do not alter $\bar{M}_n$ but they generate branched molecules, a broader molecular weight distribution and thus a larger value of the weight-average molecular weight ($\bar{M}_w$). Polymerisation with multifunctional (>2·0) monomers leads to crosslinking and, if the reaction is carried far enough, to an infinite molecular network.

The development of molecular weight in emulsion polymerisation is different than in bulk, solution and suspension polymerisations for several reasons. First, if a batch reactor is used, the polymer concentration at the reaction site in non-emulsion systems progresses from zero or a very low value to the amount present when the reaction is stopped. Such is not the case in emulsion polymerisation if the polymerisation takes place in the monomer-swollen polymer particles. In these systems the polymer : monomer ratio may be as high as 1 : 1 or 1 : 2 even at low values of overall conversion. Hence the probability for branching and crosslinking reactions is usually higher for emulsion processes than for bulk, solution or suspension systems. The addition of chain transfer agents is sometimes used to reduce the magnitude of this problem.

Differences in molecular weight development among various reactor types can occur for the same reasons. Thus, if a monomer-add semi-batch reactor is operated in a monomer-starved mode, the possibilities for polymer transfer and/or crosslinking reactions are enhanced. Hence one should expect large $\bar{M}_w/\bar{M}_n$ values in the products from such reactors. In addition, the high viscosity within the monomer-starved particles reduces the rate of termination and higher molecular weight can result, even though the propagation reaction is also slowed due to the small monomer concentration. These effects can be countered, at least in part, by the addition of chain transfer agent.

Molecular weight of the products from a continuous system will depend on the conditions at the reaction sites within each of the reactors that comprise the continuous system. If a large number of series-connected CSTRs are used, the conversion-space time profile is similar to the conversion-time profile of a batch reactor. Hence the molecular weights should not be too different from that of batch product. There can be some difference, however, and control of cross-linking seems to be more of a problem in continuous systems.

The conversion of monomer to polymer takes place under fixed conditions in each reactor of a CSTR series system. If only a few (two to five) reactors are used the average polymer concentration in the reaction sites will be higher for continuous reactor systems than for batch reactors. This will generally lead to more branching and cross-linking. Again this effect might be countered by modifiers and chain transfer agents. Such reagents can be added with the feed streams to the first reactor or to each of the reactors separately.

In summary, molecular weight development can be quite different among different reactor types. Hence recipe and operational (e.g. stopping the reaction at a lower conversion) variations may be necessary to produce equivalent product, especially if branching or cross-linking reactions are important.

4.8. Fouling and Coagulum Formation

Control of coagulum formation and the fouling of reactor surfaces can be major problems in emulsion polymerisation processes. Two mechanisms have been suggested for coagulum formation: (1) failure of latex stability, and (2) polymerisation in other sites such as large monomer drops or monomer pools. Coagulum formation can be influenced by recipe formulation, reactor design and reactor operation procedures. Coagulum can also be formed in the processing steps which are downstream from the reactor.

It is beyond the scope of this chapter to review, in any depth, coagulum or fouling problems. In fact, there is not an extensive literature on this subject. The recent review article by Vanderhoff[29] is recommended for those who must deal with such problems. The intent of this chapter is to discuss some differences in coagulum problems in different reactor processes.

Both batch and semi-batch reactor systems suffer from the problem of wall polymer caused, at least in part, by the repeated filling and discharging of the reactor. Latex and monomer can adhere to the wall or cooling surfaces and result in rather firmly bound wall polymer. If this polymer is not removed between batches it can absorb monomer and grow via polymerisation. This is a particularly important problem with some of the elastomers (SBR and polychloroprene) because the deposits seem to grow in an accelerated manner.

This cause of wall fouling can be avoided in a continuous system if the reactors are operated completely full. A Bayer patent describes one such reactor design for the continuous emulsion polymerisation of chloroprene.

A lack of latex stability, as mentioned above, is one reason for the formation of coagulum. Flocculation can be caused by the addition of electrolytes and/or polymeric flocculating agents. This phenomenon can be different among reactor processes. In a batch reactor all ingredients are charged and mixed before the latex is formed (unless a seed latex is used). Hence a lack of stability will not be caused by entering streams. Such is not the case with semi-batch or continuous systems, in which reagent streams are added to a reacting latex.

Some recipe components, such as buffers and initiators, are electrolytes and will cause flocculation if they are not charged correctly. Streams containing such ingredients should be as dilute as possible and they should be added at a point in the reactor where they will be mixed rapidly. Care should also be exercised with the addition of emulsifiers and stabilisers, especially polymeric stabilisers which can serve to stabilise or flocculate the latex. Rapid mixing is again desirable.

Coagulum and fouling can be more significant problems in continuous processes. If necessary, batch and semi-batch reactors can be cleaned after each reaction cycle. Most continuous processes, however, need to operate for rather long periods of time if they are to be economical. A continuous process which must be shut down for cleaning frequently is not continuous. Hence special attention should be given to potential flocculation and fouling problems during development of continuous processes.

5. SUMMARY

Emulsion polymerisation is a complex heterogeneous reaction system which is used to produce commercial products in a wide variety of reactor processes. The nature of the chemical and physical mechanisms involved is reasonably well understood from a qualitative point of view. The quantitative prediction of polymerisation kinetics and product properties, however, is not possible for most commercial recipes without some experimental effort. The state of theoretical knowledge is sufficient to help with the intelligent planning of product and process development programmes.

As more work is reported, the ability to custom-tailor emulsion polymers for specific applications will be enhanced. Likewise, the capability to design commercial processes to produce these latex products is improving rapidly. Many exciting problems will continue to arise in the emulsion polymerisation arena because new uses and new technology concepts are introduced frequently. The rapid expansion of the literature that has been characteristic of the 1970s should continue through the 1980s.

REFERENCES

1. Hamielec, A. E. and MacGregor, J. F., Ch. 9 in *Emulsion polymerization*, I. Piirma (Ed.), Academic Press, New York (1982).
2. Sundberg, D. C., Hsieh, J. Y., Soh, S. K. and Baldus, R. F., Paper 20 in *Emulsion polymers and emulsion polymerization*, D. Bassett and A. Hamielec (Eds), ACS Symposium Series 165, Washington, DC (1981).
3. Ugelstad, J., El Aasser, M. S. and Vanderhoff, J. W., *J. Polym. Sci., Polym. Lett. Ed.*, **11**, 505 (1973).
4. Durbin, D. P., El Aasser, M. S., Vanderhoff, J. W. and Poehlein, G. W., *J. Appl. Polym. Sci.*, **24**, 703 (1979).
5. Morton, M., Kaizerman, S. and Altier, M. W., *J. Colloid Sci.*, **9**, 300 (1954).
6. Tseng, C. M., El Aasser, M. S. and Vanderhoff, J. W., Paper 11 in *Computer applications in applied polymer science*, T. Provder (Ed.), ACS Symposium Series 197, Washington, DC (1982). ●
7. Ugelstad, J., Mørk, P. C., Berge, A., Ellingsen, T. and Khan, A. A., Ch. 11 in *Emulsion polymerization*, I. Piirma (Ed.), Academic Press, New York (1982).
8. Ugelstad, J. *et al.*, In *Science and technology of polymer colloids*, NATO Adv. Study Institute Series (Vol. 1), G. Poehlein, R. Ottewill and J. Goodwin (Eds), Martinus Nijhoff Publishers, The Hague (1983), pp. 51–99.
9. Smith, W. V. and Ewart, R. H., *J. Chem. Phys.*, **16**, 592 (1948).
10. Gershberg, D. B. and Longfield, J. E., 54th Annual AIChE Mtg, Preprint No. 10, Cleveland (1961).
11. Gerrens, H. and Kuchner, K., *Br. Polym. J.*, **2**, 18 (1970).
12. DeGraff, A. W. and Poehlein, G. W., *J. Polym. Sci.*, **A-2** (9) 1955 (1971).

13. Roe, C. P., *Ind. Eng. Chem.*, **60,** 20 (1968).
14. Priest, W. J., *J. Phys. Chem.*, **56,** 1077 (1952).
15. Brooks, B. W., *Br. Polym. J.*, **5,** 199 (1973).
16. Fitch, R. M., Prenosil, M. B. and Sprick, K. J., *J. Polym. Sci.*, **C27,** 95 (1969).
17. Fitch, R. M. and Tsai, C. H., In *Polymer colloids*, R. M. Fitch (Ed.), Plenum Press, New York (1971), p. 73 and p. 103.
18. Fitch, R. M., Paper 1 in *Emulsion polymers and emulsion polymerization*, D. Bassett and A. Hamielec (Eds), ACS Symposium Series 165, Washington, DC (1981).
19. Hansen, F. K. and Ugelstad J., *J. Polym. Sci., Polym. Chem. Ed.*, **16,** 1953 (1978); **17,** 3033 (1979); **17,** 3047 (1979).
20. Hansen, F. K. and Ugelstad, J., *J. Polym. Sci., Polym. Chem. Ed.*, **17,** 3069, (1979).
21. O'Toole, J. T., *J. Appl. Polym. Sci.*, **9,** 1291 (1965).
22. Ugelstad, J., Mørk, P. C. and Aasen, J. O., *J. Polym. Sci.*, **A-1** (5) 2281 (1967).
23. Poehlein, G. W., Dulner, W. and Lee, H. C., *Br. Polym. J.*, **14,** 153 (1982).
24. Gilbert, R. G. and Napper, D. H., *J.C.S. Faraday I*, **70,** 391 (1974).
25. Hawkett, B. S., Napper, D. H. and Gilbert, R. G., Ibid., **71,** 2288 (1975); **73,** 690 (1977); **76,** 1323 (1980); **76,** 1344 (1980).
26. German Patent 2,520,891, Bayer AG (1976).
27. Bassett, D. R. and Hoy, K. L. Paper 23 in *Emulsion polymers and emulsion polymerization*, D. Bassett and A. Hamielec (Eds), ACS Symposium Series 165, Washington, DC (1981).
28. Johnston, J. A., Bassett, D. R. and MacRury, T. B., Paper 24, Ibid.
29. Vanderhoff, J. W., In *Science and technology of polymer colloids*, NATO Adv. Study Inst. Series (Vol. 1), G. Poehlein, R. Ottewill and J. Goodwin (Eds), Martinus Nijhoff Pub., The Hague (1983), pp. 1–39.

Chapter 3

Adsorption from Solution—
Part I: Low Molecular Weight,
Ionic Adsorbates

R. O. JAMES

*Emulsion Research Laboratories, Eastman Kodak Company, Rochester,
New York, USA*

1. INTRODUCTION

Polymer colloids or latices are interesting materials both practically
and theoretically. They have been widely used in paints, adhesives,
papermaking, and medical science and technology because, for example, they may be film-forming, they may adsorb and transport ionic or
molecular species, or they may modify the optical and mechanical
properties of fibres. In recent years, monodisperse latices have been
useful in testing theories of colloid behaviour because of their simple
geometry and controlled surface composition.

The utility of latices and many other colloid systems is governed by
two principal properties. First, the bulk properties of the material must
be suitable, and, second, the dispersion of particles must remain stable
during the time before its application. That is, the particles should not
form aggregates that would separate either in the container or on the
material to which they were applied.

In general, the ever-present attractive van der Waals or dispersion
forces between particles that cause adhesion and aggregation are
controlled by the physico-chemical properties of the bulk material of
the components of the dispersion. These properties include composition and size of the particles, and composition and polarity of the
dispersing liquid. On the other hand, the surface structure, which may

include adsorbed layers of solvent, ions, molecules or polymers, controls the repulsive interactions between particles by either electrostatic or steric forces. Thus the surface properties allow us, by chemical means, to control the colloidal stability of a latex that would otherwise form aggregates as the particles collided during Brownian motion. There are essentially two ways of controlling colloidal stability. The repulsive barrier between particles may be due to either the adsorption of small ionic species, which causes an electrostatic repulsion, or the adsorption of layers of large, solvated polymers. Many latices are prepared with functional groups (e.g. carboxylic acid, sulphonic acid or alkyl amine) on the surface by polymerisation or adsorption from the solution. The adsorption of ions and solvent usually leads to dissociation or reaction of these functional groups, which then form a charge layer on the surface. This causes an accumulation of opposite charge in the solution adjacent to the interface and a decay of electrostatic potential from the surface into the solution. The arrangement of charges and the associated electrostatic potentials are termed the electrical double layer (EDL). When colloids are stabilised by the mutual repulsion of their overlapping double layers, they are often described as being 'charge stabilised'.

When latices have adsorbed or anchored polymer chains, colloidal stability is enhanced owing to the effective repulsion arising from interpenetration and compression of the chains. The polymers may be ionic (e.g. gelatin) or non-ionic (e.g. poly(ethylene glycol) and poly(vinyl alcohol)). Colloidal particles stabilised by polymers or 'protective colloids' are said to be 'sterically stabilised'. An advantage of this method of stabilisation is that it is much less sensitive to the presence of other ionic solutes. The adsorption of polymers will be discussed in Chapter 4.

For the moment, we shall deal with those latices whose surfaces contain bound or adsorbed ions and molecules that may contain ionisable functional groups.

2. ADSORPTION MODELS

Theories or models of adsorption are developed so that the amount of a solute associated with an interface (Γ_i, mol m^{-2}) can be related to the amount of solute in the solution (C_i, mol dm^{-3}) by means of an adsorption isotherm.

The model should be constructed so that the adsorption reaction can be characterised by a number of parameters. These include:

(1) A parameter that measures the tendency for adsorption to occur, i.e. how strongly the solute associates with the surface. This is the free energy of adsorption ΔG_{ads} or the related equilibrium constant K_{ads}.

(2) A parameter that describes how much solute can be adsorbed. Sometimes there may be no apparent limit to adsorption; however, for latices there is often a limit that corresponds to the number of ionisable groups or perhaps the number of molecules that can be packed in an adsorbed monolayer.

(3) The stoichiometry that summarises the ions and molecules participating in adsorption reactions. For example, an ion may simply adsorb at an interface, perhaps displacing solvent, or it may displace another ion that diffuses into the solution.

(4) Equations and parameters that allow for the formation of an electrical double layer and that describe the structure or charge-potential relationships in the EDL.

Not all theories or models contain all of these elements. Generally, when more of these elements are included, obtaining a mathematical solution is more difficult. Despite this, there has been a rapid increase in the use of the more detailed adsorption models.

2.1. Langmuir Isotherm

One isotherm that has proved useful in many situations, including adsorption of gases and solutes, is the Langmuir isotherm.[1] This isotherm, which can be deduced from either kinetic[2] or thermodynamic[2,3] reasoning, includes the concepts of adsorption tendency, stoichiometry, molecular dimensions or adsorption capacity, and mass balance of reactants. As will be seen, the isotherm can be extended to include ionic terms, which allows it to be applied in EDL models. The thermodynamic derivation of the model given by Parks[3] will be followed here.

Imagine the adsorption of solute X at the latex/water interface, where S is an adsorption site and SX is an occupied site. During uptake, some adsorbed water will be displaced by X. If it is assumed that one water molecule is displaced by each solute X molecule, the mass-action reaction can be written as:

$$X_{aqueous} + S(H_2O) \rightleftharpoons SX + H_2O \tag{1}$$

The tendency to adsorb is expressed by an equilibrium constant K_{ads} in terms of reactants and products:

$$K_{ads} = a_W a_{SX}/(a_X a_{SW})$$

where K_{ads} is dimensionless, and the related free energy of adsorption, ΔG_{ads}, is defined as:

$$\Delta G_{ads} = -RT \ln K_{ads} \tag{2}$$

To evaluate K_{ads} and ΔG_{ads} from experimentally measurable quantities, the activities of the components a_i must be related to their concentrations n_i by proportionality constants or activity coefficients k_i, e.g:

$$a_{SX} = k_{SX} n_{SX} = \text{activity of adsorbed X}$$
$$a_{SW} = k_{SW} n_{SW} = \text{activity of adsorbed water}$$
$$a_X = k_X n_X = \text{activity of solute X}$$
$$a_W = k_W n_W = k_W \frac{N_a \rho}{M} = \text{activity of water in aqueous phase}$$

where n_w = number concentration of water molecules, dm^{-3}
n_X = number concentration of solute molecules, dm^{-3}
n_{SX} = surface number concentration of adsorbed solute molecules, m^{-2}
n_{SW} = surface number concentration of sites unoccupied by X, molecules m^{-2}
N_a = Avogadro's number
ρ = density of water, $g\,dm^{-3}$
M = molecular weight of water, g

A further simplifying assumption is that the activity-coefficient terms in the mass law (eqn (1)) are constant and cancel. Then, if the total number of sites per unit area is n_{max}:

$$n_{max} = n_{SW} + n_{SW} \tag{3}$$

and

$$K = n_W n_{SX}/n_X(n_{max} - n_{SX}) \tag{4}$$

This may be rearranged in the form of an isotherm relating the adsorption density n_{SX} to the solute concentration, viz:

$$n_{SX} = \frac{n_{max}K_{ads}n_X/n_W}{1 + K_{ads}n_X/n_W} = \frac{n_{max}K_{ads}(n_X/N_a)(M/\rho)}{1 + K_{ads}(n_X/N_a)(M/\rho)}$$

$$= \frac{n_{max}}{1 + (\rho/M)(N_a/n_X)/K_{ads}} \tag{5}$$

Sometimes, for convenience, the constant term $M/N_a\rho$ is included in the adsorption constant, so that if $K' = KM/N_a\rho$:

$$n_{SX} = n_{max}K'n_X/(1 + K'n_X) \tag{6}$$

or if the concentration of the dilute solute X is expressed as a mole fraction, $X \simeq (n_X/N_a)(M/\rho)$, then:

$$n_{SX} = n_{max}KX/(1 + KX) \tag{7}$$

These equations are all variations of the Langmuir isotherm. Clearly, for low solute concentrations $n_{SX} \simeq n_{max}KX$, but for high solute concentrations n_{SX} approaches the limiting value n_{max}.

When isotherms obey the form of the Langmuir equation, the saturation coverage or site density n_{max} and the binding constant K are easily evaluated by plotting the inverse form $1/n_{SX}$ against $1/X$, e.g:

$$1/n_{SX} = (1/Kn_{max})/X + 1/n_{max} \tag{8}$$

Although this isotherm provides a basis for interpretation of adsorption of molecular and ionic solutes, the model on which it is based is too restrictive to apply to a wide range of ionic systems.[4] The implicit use of a binding constant, which reflects the interaction of an adsorbate with only one site, disallows the real possibility that the binding energy may vary strongly as the site coverage changes owing to a change of surface charge and electrostatic interactions, and changes in the packing or orientation of adsorbed molecules. However, if it is assumed that the binding parameter $K = \exp(-\Delta G_{ads}/RT)$ may vary with such changes, terms can be incorporated to account for these additional interactions that occur during adsorption.

2.2. The Gouy–Chapman Model for the Diffuse Electrical Double Layer

In the Gouy–Chapman (GC) model of the interfacial region, the electrical forces determining the uptake of ions in the liquid adjacent to the charged solid are opposed by thermal diffusion forces. A

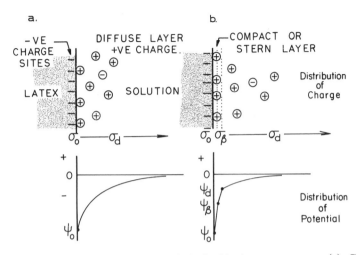

Fig. 1. Schematic diagram of electrical double layer structure. (a) Gouy–Chapman diffuse layer, and (b) Gouy–Chapman–Stern–Grahame compact and diffuse double layers showing a negatively charged surface and a net positive charge in solution.

schematic representation of the model is shown in Fig. 1. Extensive discussion of the theory for this model, developed by Gouy[5] and Chapman,[6] is available in several texts (see Hunter[7] and Hiemenz[8]) and reviews (see Grahame[9]). The model was derived on the basis of three equations in classical electrostatic theory which describe the rate of change of the electric field near an electrostatic charge. The Poisson equation in Cartesian coordinates can be written as:

$$\nabla^2 \psi = -\rho/\varepsilon \tag{9}$$

where ε is the permittivity of the medium (equal to the dielectric constant of the medium, D, times the permittivity of free space $\varepsilon_0 = 8\cdot854 \times 10^{-12}$ F m^{-1}), ψ is the potential and ρ is the volume charge density.

The charge density in some volume in the liquid is given by:

$$\rho = \sum_{\text{all ions}} n_i z_i e \tag{10}$$

where n_i is the concentration of ion i, z_i is the sign and magnitude of charge on i, and e is the magnitude of the electronic charge.

The distribution of ions in the electric field perturbed by thermal diffusion is calculated from the Boltzmann equation:

$$n_i = n_i^{\circ} \exp\left(-W_i/\mathbf{k}T\right) \tag{11}$$

where n_i° is the bulk solution concentration, W_i is the work or free energy of adsorption, and $\mathbf{k}$ is Boltzmann's constant. It is assumed that there is only electrical work involved so that:

$$W_i = z_i e \psi \tag{12}$$

No other energetic interactions, e.g. solvation changes or short-range binding, are considered.

These equations are combined to give the Poisson–Boltzmann equation, here written for a flat plate surface:

$$\frac{d^2\psi}{dx^2} = -\frac{1}{\varepsilon} \sum_i n_i^{\circ} z_i e \exp\left(-z_i e\psi/\mathbf{k}T\right) \tag{13}$$

Using the boundary conditions that the potential at the innermost part of the diffuse EDL is given by $\psi_0 = \psi_d$ at $x = 0$ and at $x = \infty$ by $\psi = 0$, and a form of Gauss' law relating the electric field strength to the charge density of the diffuse layer σ_d and the surface charge σ_0:

$$\frac{d\psi_0}{dx} = -\frac{\sigma_0}{\varepsilon} = \frac{+\sigma_d}{\varepsilon} \tag{14}$$

one may obtain a number of useful equations:

$$\sigma_0 = \left\{2\varepsilon_0 D\mathbf{k}T \sum n_i^{\circ}[\exp\left(-ze\psi_d/\mathbf{k}T\right) - 1]\right\}^{\frac{1}{2}} \tag{15}$$

For symmetrical electrolytes ($|z_+| = |z_-|$)

$$\sigma_d = -(4n^{\circ}ze/\kappa) \sinh\left(ze\psi_d/2\mathbf{k}T\right) \tag{16}$$

$$= -0\cdot1174\sqrt{(C)} \sinh\left(ze\psi_d/2\mathbf{k}T\right) \quad \text{C m}^{-2} \tag{17}$$

where for water at 25°C:

$$\kappa = \left(\frac{e^2 \sum n_i^{\circ} z_i^2}{\varepsilon_0 D\mathbf{k}T}\right)^{\frac{1}{2}} \tag{18}$$

$$= \left(\frac{2000F^2}{\varepsilon_0 DRT}\right)^{\frac{1}{2}}\sqrt{I} \quad \text{m}^{-1} \tag{19}$$

$$= 3\cdot288\sqrt{I} \quad \text{nm}^{-1} \tag{20}$$

where F is the Faraday constant and the ionic strength is given as $I = \frac{1}{2}\sum C_i z_i^2$, where C_i is the molar concentration of each ion.

The term $1/\kappa$ is an effective thickness parameter measuring the decay of the diffuse double layer into the solution.

Recall that the adsorption density of an individual ion i, Γ_{G_i}, in the Gouy–Chapman diffuse layer is the sum of the excess interfacial concentration, i.e:

$$\Gamma_{G_i} = \int_{x=d}^{\infty} (n_{i,x} - n_{i,\infty})\, dx$$

then the integrated form is:[3,9]

$$\Gamma_{G_i} = \left(\frac{2000\varepsilon RT}{z_i^2 F^2}\right)^{\frac{1}{2}} \sqrt{(C_i)}[\exp(-zF\psi_d/2RT) - 1] \quad (21)$$

or

$$\sigma_{d,i} = z_i F\Gamma_{G_i} = 0\cdot0587\sqrt{(C_i)}[\exp(-19\cdot46z_i\psi_d) - 1] \quad (22)$$

where, for aqueous solutions at 25°C, σ_d is in $C\,m^{-2}$, C is in $mol\,dm^{-3}$, and ψ_d is in V. The adsorption density of minor solutes Γ_j can also be written in terms of the adsorption density of the major solutes and their bulk concentrations, e.g:

$$\Gamma_j = \frac{C_j}{C_i}\Gamma_i \quad (23)$$

Another quantity that can be obtained from the theory and that is measurable in some electrochemical experiments is the capacitance of the diffuse double layer C_d

$$C_d = -\frac{d\sigma_d}{d\psi_d} = \varepsilon\kappa \cosh(ze\psi_d/2\mathbf{k}T) \quad F\,m^{-2} \quad (24)$$

$$= 2\cdot285\, z\sqrt{(C)} \cosh(ze\psi_d/2\mathbf{k}T), \quad F\,m^{-2} \quad (25)$$

if C is in $mol\,dm^{-3}$ and the solvent is water at 25°C.

The theory for flat plates can be applied to spheres for conditions where the radius a of a particle is large compared to the thickness $(1/\kappa)$ of the double layer, i.e. $\kappa a \geqslant 10$. For small particles or very dilute electrolytes, there is no exact analytic solution of the Poisson–Boltzmann equation[7,10] (spherical coordinates must be used, however). Consequently, there have been many efforts to obtain reliable approximate analytic solutions of the Poisson–Boltzmann equation for a spherical colloidal particle.[7,11,12] The simplest approximation for the spherical particle is to use the linearised Debye–Hückel approximation.

For a particle of radius a, surface potential ψ_d, surface charge $Q = 4\pi a^2 \sigma_0$, where σ_0 is the surface charge density and σ_d is the diffuse layer charge, the following relations for the double layer potential ψ at radial distance r from the particle centre may be obtained[7] when $\psi_0 < 25$ mV:

$$\psi = \psi_0 \frac{a}{r} \exp\left[-\kappa(r-a)\right] \tag{26}$$

$$\sigma_0 = \frac{Q}{4\pi a^2} = \varepsilon \frac{(1+\kappa a)}{a} \psi_0 \tag{27}$$

and

$$\psi = -\sigma_0 \frac{a^2}{\varepsilon r(1+\kappa a)} \exp\left[-\kappa(r-a)\right] \tag{28}$$

For $\kappa a < 0.1$, eqn (27) is good for $\psi_0 < 100$ mV. Another relation derived in a 'zero-order approximation' by White[11] relates the surface charge density of the sphere to that of a flat plate. It is claimed that this provides a better approximation to the exact numerical solution than eqn (28) for $\kappa a > 2$.

$$\sigma_{\text{sphere}} = \left(\frac{\kappa a + 1}{\kappa a}\right)\sigma_{\text{plate}} \tag{29}$$

$$= -0.1174\sqrt{(C)}\left(\frac{\kappa a + 1}{\kappa a}\right)\sinh\left(ze\psi_d/2\mathbf{k}T\right) \quad \text{C m}^{-2} \tag{30}$$

More recently, Ohshima et al.[12] gave a further improved approximate result for univalent electrolytes:

$$\sigma_{\text{sphere}} = \varepsilon\kappa \frac{2\mathbf{k}T}{e} \sinh\left(e\psi/2\mathbf{k}T\right)$$

$$\times \left\{1 + \frac{2}{\kappa a \cosh^2\left(e\psi/4\mathbf{k}T\right)} + \frac{8\log_e[\cosh\left(e\psi/4\mathbf{k}T\right)]}{\kappa^2 a^2 \sinh^2\left(e\psi/2\mathbf{k}T\right)}\right\}^{\frac{1}{2}} \tag{31}$$

The Gouy–Chapman theory produces qualitative agreement with observed behaviour for many systems including latices. For example, the surface charge density increases with salt concentration, whereas the magnitude of the surface potential decreases with salt concentration. At low ionic strengths the interfacial capacitance has a minimum at the point of zero charge. For poly(vinyl toluene) latices with low surface charge density (e.g. $0 \cdot 5 \ \mu C \ cm^{-2}$) due to ionised alkyl sulphate functional groups, Bagchi et al.[13] observed excellent agreement between the surface charge density obtained by titration of surface groups and that obtained by using the measured zeta potential, estimated using measurements of the migration velocity of particles in an electric field, in eqn (17). This potential is the electrostatic potential at the hydrodynamic shear plane near the particle surface, but in the electrical double layer.

For latices that have higher surface charge density or more ionisable functional groups, the model does not perform as well. For example, the surface charge density σ_0 from titration measurements and the diffuse-layer charge density obtained from the zeta potential σ_d do not balance.[14,15] This implies that counterions are located in a compact charge layer between the surface and the hydrodynamic shear plane. It is also possible that ions in a compact adsorbed layer may experience short-range 'chemical binding interactions' with the surface.

One way to account for this possibility was proposed by Stern[16] and extended by Grahame.[9] They developed a model, usually called the Gouy–Chapman–Stern–Grahame (GCSG) model, which combined the features of the Langmuir and GC models and has proved useful for many colloid systems.

2.3. Gouy–Chapman–Stern–Grahame (GCSG) Model for the EDL of Compact and Diffuse Layers

In a number of interfacial studies, certain ions were found that shifted the potential of the point of zero charge, whereas others were capable of completely reversing the zeta potential. This behaviour is not described by the GC theory because it considers only electrostatic driving forces in adsorption processes. Stern[16] and later Grahame[9] proposed models that allowed ions in the innermost part of the EDL, between the surface and the hydrodynamic shear plane, to participate in specific, non-Coulombic, short-range interactions at the surface. This layer, called the Stern layer, contains charge (σ_β, $C \ m^{-2}$) so that the electroneutrality equation relating the surface charge σ_0, Stern-layer charge σ_β, and diffuse-layer charge σ_d becomes:

$$\sigma_0 + \sigma_\beta + \sigma_d = 0 \tag{32}$$

The distribution of charges is then viewed as a triple layer, as shown in Fig. 1(b). These layers of charge are analogous to a molecular condenser in the interfacial region. Where the surface charge develops from binding reactions at specific sites, as may occur for latices, the models are also known as site-binding or surface-complexation models. The isotherm is based on the Langmuir isotherm, where Γ_{max} is the total site density for ions of charge z_i, and the free energy of adsorption ΔG_i is composed of a Coulombic contribution and a specific adsorption contribution. The charge and adsorption density of i are given by:

$$\sigma_{\beta i} = z_i F \Gamma_i = \frac{z_i F \Gamma_{max} K_i X_i}{1 + \sum K_i X_i} \quad \text{C m}^{-2} \tag{33}$$

If univalent cations and anions are adsorbed in the Stern layer, the charge is given by:

$$\sigma_\beta = \sigma_{\beta+} + \sigma_{\beta-} = \frac{F \Gamma_{max}(K_+ X_+ - K_- X_-)}{1 + K_+ X_+ + K_- X_-} \quad \text{C m}^{-2} \tag{34}$$

Under some conditions, if the specific adsorption energy Φ is approximately constant, the contribution of specific adsorption to K_i can be expressed as an intrinsic binding constant K^{int}, which is more or less independent of the surface charge. Then:

$$K_+ = \exp\left[-(\Phi_+ + z_+ e\psi_\beta)/\mathbf{k}T\right] = K_+^{int} \exp\left(-z_+ e\psi_\beta/\mathbf{k}T\right)$$
$$K_- = \exp\left[-(\Phi_- + z_- e\psi_\beta)/\mathbf{k}T\right] = K_-^{int} \exp\left(-z_- e\psi_\beta/\mathbf{k}T\right) \tag{35}$$

Although this assumption is not rigorous, it has proved useful in a number of studies of latices.[17-19]

The diffuse-layer charge density σ_d is given by eqn (17). To complete the description of the decay of electrostatic potential through the capacitor-like compact layers, the equations for parallel-plate capacitors can be written as:

$$\psi_0 = \psi_\beta + \frac{\sigma_0}{K_1} \tag{36}$$

and

$$\psi_\beta = \psi_d - \frac{\sigma_d}{K_2} \tag{37}$$

where K_1 and K_2 are the inner-layer capacitances of the compact part of the EDL (in $F\,m^{-2}$). At this point, a way is still needed to evaluate the surface charge density σ_0 in terms of the solution composition. The six unknowns σ_0, ψ_0, σ_β, ψ_β, σ_d and ψ_d can then be evaluated from the six equations, if values are assumed or determined for K_1 and K_2. For strongly acidic functional groups (e.g. $ROSO_3H$), which are ionised under most solution conditions, the surface charge density σ_0 will be given by:

$$\sigma_0 = -FN_{ROSO_3^-}/A$$

where $N_{ROSO_3^-}$ is the number of moles of functional groups and A is the surface area of the latex. Many other colloids, however, may have weaker acid functional groups, e.g. carboxylic acids or amines. Their degree of ionisation will depend on the pH of the solution, and this leads to the study of ionisable surface models.

2.4. Ionisable Surface Models for Latices

It might be imagined that the adsorption reactions of colloidal particles occur at an array of surface sites that act as acids, bases or complex-forming ligands. Owing to ionic reactions, a surface charge and a potential difference are established. The distribution of ions in the EDL in the solution is described by either a GC or a GCSG model. The development of these models has been reviewed by Healy and White,[20] James and Parks[21] and Hunter.[7]

This family of models will now contain many of the important features that allow comparison of theoretical predictions with a range of experimental observations. They describe the relationship between the surface charge and potential, and the solution composition in terms of the density of surface sites, their ionisation constants, the electrolyte concentration and pH.

2.4.1. Simple Diffuse-Layer/Ionisable-Surface Models

A variety of types of surface may be considered including: (1) monofunctional surfaces (e.g. anionic or cationic latices[13,22] or adsorbed ionisable monolayers[23]); (2) amphoteric surfaces (e.g. metal oxide hydrosols[20,21]); and (3) zwitterionic surfaces with different, oppositely charged functional groups (e.g. latices[24–26] and proteins).

In these models, the effect of electrolyte concentration C on the surface charge σ_0 is exerted only through the concentration dependence of the diffuse-layer charge σ_d. This is in contrast to electrolyte-binding models in which the concentration contributes directly to σ_0

and σ_β. Here only the simplest case of a latex with monofunctional surface groups (e.g. —COOH) will be dealt with; more complex samples have been reviewed.[20,21]

Assume that for a weak-acid functional group —SH, the site density is S_T/A mol m^{-2}, where S_T is the concentration of sites and A is the surface area. At the surface, dissociation may occur, e.g:

$$SH \overset{K_a^{int}}{\rightleftharpoons} S^- + H_S^+ \qquad (38)$$

where $K_a^{int} = [S^-][H_S^+]/[SH]$. The interfacial concentration of protons H_S^+ can be related to the bulk solution concentration H^+ by:

$$[H_S^+] = [H^+] \exp(-e\psi_0/kT) \qquad (39)$$

The surface charge density is given by:

$$\sigma_0 = -F[S^-]/A \qquad C\,m^{-2} \qquad (40)$$

if the unit for $[S^-]$ is mol dm^{-3} and for A is m^2 dm^{-3}. The mass balance on the sites is:

$$S_T = [SH] + [S^-] \qquad (41)$$

and the degree of ionisation is measured by:

$$\alpha_- = [S^-]/([SH] + [S^-]) = \sigma_0 A/FS_T \qquad (42)$$

Equations (41) and (42) can be combined to give:

$$\sigma_0 = -F\alpha S_T/A = -FS_T/\{A(1 + [H^+]\exp(-e\psi_0/kT)/K_a^{int})\} \qquad (43)$$

Because the surface charge must be balanced by the diffuse-layer charge:

$$\sigma_0 + \sigma_d = 0 \qquad (44)$$

for $\kappa a \gg 10$ and univalent electrolytes:

$$\sigma_0 = -\sigma_d = \left(\frac{4000CF}{\kappa}\right)\sinh(e\psi_0/2kT)$$

$$= \frac{FS_T/A}{\gamma}\sinh(e\psi_0/2kT) \qquad C\,m^{-2} \qquad (45)$$

where C is the concentration of electrolyte in mol dm^{-3}, S_T/A is the site density, κ has units m^{-1}, and γ is a dimensionless parameter defined by:

$$\gamma = S_T\kappa/4000CA \qquad (46)$$

The equation obtained by elimination of σ_0, F, S_T and A from eqns (43) and (45):

$$-1/\{1+[H^+]\exp(-e\psi_0/kT)/K_a^{int}\} = \sinh(e\psi_0/2kT)/\gamma \qquad (47)$$

can be solved[20,21] for the unknown surface potential ψ_0 for given values of S_T, A, K_a^{int}, $[H^+]$ and κ. Once ψ_0 has been evaluated, α_- or σ_0 can be obtained by back-substitution into eqn (43) or eqn (45). For a particular dispersion characterised by the parameters A, S_T and K_a^{int}, and solution conditions defined by ionic strength, values can be calculated for the surface charge σ_0 and ionisation α_-, which are experimentally accessible, and the surface potential, which can be related to the zeta potential. Using this approach, Healy and White[20] calculated the effect of pH on the surface charge density of a poly(styrene) latex with about $0\cdot87$ sites nm^{-2}, which when fully ionised had a charge density of -14 $\mu C\,cm^{-2}$ for particular values of ionic strength and acidity constants pK_a^{int}. The calculated surface potential was somewhat higher than that normally observed for the zeta potential. However, by assuming various distances for the hydrodynamic shear plane, they could estimate ζ.

2.4.2. Determination of Surface Ionisation Constants

The calculations described above rely on the use of an intrinsic surface acidity constant K_a^{int}. The value of this term can be obtained from titration data by application of the model, viz:

$$K_a^{int} = [S^-][H^+]\exp(-e\psi_0/kT)/[SH] \qquad (48)$$

$$[S_{TOT}] = [SH]+[S^-] \qquad (49)$$

Elimination of the [SH] term from eqns (48) and (49) gives:

$$[H^+]\exp(-e\psi_0/kT) = K_a^{int}[S_{TOT}]/[S^-] - K_a^{int} \qquad (50)$$

Both $[H^+]$ and $[S^-]$ are available from titration data, and the electrostatic interaction term can be estimated from the inverse form of eqn (45):

$$\psi_0 = 2\frac{kT}{e}\sinh^{-1}(\sigma_0/0\cdot1174\sqrt{C}) \qquad (51)$$

where σ_0 is in $C\,m^{-2}$ and C is in $mol\,dm^{-3}$ at 25°C, or from experimental zeta-potential data. A plot of $[H^+]\exp(-e\psi_0/kT)$ against $1/[S^-]$ should yield K_a^{int} and $[S_{TOT}]$ from the slope and the intercept, respectively.[21,27]

This type of analysis of the surface-ionisation diffuse double layer model can be extended to more complicated amphoteric or zwitterionic colloids, although in these cases the relationship between the fractional ionisation $\alpha_\pm$ and the surface charge density σ_0 is not as direct.

2.4.3. Compact Double Layer/Ionisable-Surface Models

For a number of colloids, including latices and oxides, with high surface charge densities (e.g. $>5\ \mu C\,cm^{-2}$), the surface charge from titration and the diffuse-layer charge at the hydrodynamic shear plane usually do not agree. For this and other reasons a compact layer of charge is thought to be located between the surface and the shear plane, as shown in Fig. 1(b). A particular model described by James et al.[17] is outlined below, with a latex with weak-acid functional groups as an example.

If the solution contains both a salt MX and a base MOH or an acid HX, the cations H^+ and M^+ may react with anionic surface sites, e.g:

$$SH \underset{}{\overset{K_a^{int}}{\rightleftharpoons}} S^- + H_S^+ \tag{38}$$

and

$$SH + M_S^+ \underset{}{\overset{K_M^{int}}{\rightleftharpoons}} S^- M_S^+ + H_S^+ \tag{52}$$

The intrinsic binding constants are:

$$K_a^{int} = [S^-][H^+] \exp(-e\psi_0/kT)/[SH] \tag{53}$$

and

$$K_M^{int} = [SM][H^+] \exp(-e\psi_0/kT)/\{[SH][M] \exp(-e\psi_\beta/kT)\} \tag{54}$$

where ψ_β is the mean electrostatic potential in the layer containing the ion M^+, so that $e\psi_\beta$ is the electrical part of the work of adsorption.

The surface charge density evaluated by potentiometric titration is due to all reactions involving proton or hydroxyl transfer at the surface, so that:

$$\sigma_0 = F(\Gamma_{H^+} - \Gamma_{OH^-}) = F(-[S^-] - [S^- M^+])/A$$
$$= F([S_{TOT}] - [SH])/A \tag{55}$$

The Stern-layer or compact-layer charge density is due to bound cations M^+, thus:

$$\sigma_\beta = F[S^- M^+]/A \qquad C\,m^{-2} \tag{56}$$

The diffuse-layer charge density σ_d is given by eqn (17) for $\kappa a > 10$ or an approximate form like eqn (31) for spherical particles where $\kappa a \lesssim 1$. By use of the diffuse-layer potential ψ_d for $\kappa a \gtrsim 10$ and univalent electrolytes:

$$\sigma_d = -0 \cdot 1174\sqrt{(C)} \sinh (e\psi_d/2\mathbf{k}T) \qquad \mathrm{C \, m^{-2}} \qquad (17)$$

These charge layers are linked by the electroneutrality equation:

$$\sigma_0 + \sigma_\beta + \sigma_d = 0 \qquad (57)$$

For small particles, this relation should be based on the total charges Q_0, Q_β, and Q_d in each layer rather than the charge density.[7]

For a parallel-plate molecular condenser the charge-potential relationships in the compact layer can be written as:

$$\psi_0 = \psi_\beta + (\sigma_0/K_1) \qquad (58)$$

and

$$\psi_\beta = \psi_d - (\sigma_d/K_2) \qquad (59)$$

where the diffuse-layer potential ψ_d is often identified with the experimental zeta potential ζ.[28]

The mass balance on the surface sites is given by:

$$[S_T] = [SH] + [S^-] + [SM] \qquad (60)$$

This set of simultaneous equations can be solved for ψ_0, σ_0, ψ_β, σ_β, ψ_d and σ_d, if values for S_T, K_a^{int}, K_M^{int}, electrolyte concentrations C_{MX}, C_{MOH}, and C_{HX}, and capacitance values K_1 and K_2 are known.[17,21] The computer programs MINEQL (FORTRAN) and MICROQL (BASIC) written by Westall and co-workers[29,30] are available for the solution of such problems.

2.4.4. Estimation of Surface Binding Constants

The values of binding parameters K_a^{int} and K_M^{int} can be estimated by considering the effect of surface charge and ionic strength on the experimentally accessible acidity quotients Q_a, where:

$$K_a^{int} = Q_a \exp (-e\psi_0/\mathbf{k}T) = [S^-][H^+] \exp (-e\psi_0/\mathbf{k}T)/[SH] \qquad (61)$$

Q_a can be evaluated at each titration data point from S_T and $[H^+]$. By use of the mass balance on sites (eqn (60)) and expressions for K_a^{int} and K_M^{int}, it can be shown that:

$$[S^-] = ([S_T] - [SH])/(1 + K_M^{int}[M^+] \exp (-e\psi_\beta/\mathbf{k}T)/K_a^{int}) \qquad (62)$$

and substitution into eqn (61) gives:

$$K_a^{int} = ([S_T] - [SH])[H^+] \exp(-e\psi_0/kT)/\{[SH]$$
$$\times (1 + K_M^{int}[M^+] \exp(-e\psi_\beta/kT)/K_a^{int})\} \tag{63}$$

Because $\sigma_0 = -F([S_T] - [SH])/A$ and $\sigma_{max} = -F[S_T]/A$:

$$K_a^{int} = \left(\frac{\sigma_0}{\sigma_{max} - \sigma_0}\right) \frac{[H^+] \exp(-e\psi_0/kT)}{\{1 + K_M^{int}[M] \exp(-e\psi_\beta/kT)/K_a^{int}\}} \tag{64}$$

or

$$pK_a^{int} = pH - \log_{10}\frac{\alpha_-}{1 - \alpha_-} + \frac{e\psi_0}{2 \cdot 3kT}$$
$$+ \log_{10}(1 + K_M^{int}[M] \exp(-e\psi_\beta/kT)K_a^{int}) \tag{65}$$

This equation shows that $pQ_a = pH - \log \alpha_-/(1 - \alpha_-)$ varies with both the potential ψ_0 and the electrolyte concentration [M], and that $pQ_a = pK_a^{int}$ for the hypothetical condition of $\sigma_0 = f(\psi_0) = 0$ and infinite dilution. This hypothetical condition can be obtained by a double extrapolation technique when pQ is plotted against the semi-empirical function $(\alpha_- + \sqrt{[M]})$. Extrapolation of lines of constant ionic strength or constant fractional ionisation to zero gives pK_a^{int} as the intercept.[17,21]

Similarly, the intrinsic binding constant pK_M^{int} can be estimated from the logarithmic form of eqn (54):

$$pK_M^{int} = pH - \log_{10}\frac{[SM]}{[SH]} + \log_{10}[M] + \frac{e(\psi_0 - \psi_\beta)}{2 \cdot 3kT} \tag{66}$$

If one assumes that at high ionic strength most of the deprotonated sites also bind the counterion, then $\sigma_0 \simeq F[SM]/A$, and defining $\alpha_- = \sigma_0/\sigma_{max}$ gives:

$$pK_M^{int} = pH - \log_{10}\frac{\alpha_-}{1 - \alpha_-} + \log_{10}[M^+] + \frac{e(\psi_0 - \psi_\beta)}{2 \cdot 3kT} \tag{67}$$

Again, if $pQ = pH - \log[\alpha_-/(1 - \alpha)]$ is plotted against another semi-empirical term $(\alpha + \log[M^+])$, pK_M^{int} is obtained as the intercept value of pQ for the hypothetical condition that $\sigma_0 = f(\psi_0 - \psi_\beta) = 0$ and $\log[M] = 0$.[17,21] Because the inner-layer capacity of the EDL K_1 is $\sigma_0/(\psi_0 - \psi_\beta)$, this quantity can be evaluated from the slopes of such plots.[31]

2.5. Adsorption of Complex Solutes

In many applications of latices, complex solutes can be added to improve the stability of dispersions or perhaps to coagulate the dispersion to facilitate solid/liquid separations. One of the main groups of complex solutes includes anionic, non-ionic and cationic surfactants, and another includes salts of highly charged, hydrolysable metal ions. In addition to the ionic interactions with surfaces, these complex solutes participate in other kinds of interaction that promote their adsorption. The hydrocarbon chain length is an important factor for surfactants, and the tendency to hydrolyse and the solution pH are dominant factors for complex metal ions.

2.5.1. Adsorption of Surfactants

Surfactant properties are dominated by their amphipathic nature. The presence of hydrophobic and hydrophilic functional groups in a single molecule controls their solubility and their tendency to adsorb at the solid/liquid interface. Numerous studies of the physico-chemical properties of surfactant solutions have shown that at a particular concentration, the critical micelle concentration (CMC), the hydrocarbon chains associate to form stable three-dimensional clusters or aggregates called micelles. The principal driving force for this phenomenon arises from the so-called hydrophobic bonding energy, which is the free-energy change associated with the transfer of the methylene groups ($-CH_2-$) of the hydrocarbon part of the surfactant from an aqueous environment to an oil-like environment in the interior of the micelle. This arises from an entropy increase as water molecules that solvated the hydrocarbon chain are released. The free-energy contribution is about $-2/3kT$ per CH_2 for the formation of aqueous ionic micelles and about $-4/3kT$ per CH_2 for extraction of hydrocarbon from water to hydrocarbon liquid.

Adsorption of surfactants is influenced by similar effects. Indeed, for a large range of surfaces and surfactants, two-dimensional, interfacial aggregates tend to form at concentrations lower than the CMC. This increases the adsorption density in the compact part of the EDL above that expected on the basis of ionic interactions. The contribution to the free energy of adsorption due to this hydrophobic interaction, sometimes called hemi-micelle formation, is $\sim -0{\cdot}7$ to $-1{\cdot}0kT$ per methylene group. The effectiveness of this contribution is somewhat dependent on surface coverage because the possibility of two-dimensional aggregate formation increases as the fractional coverage increases. These concepts have been summarised in more detail by Healy and co-workers.[32-34] This range of values for the specific adsorption potentials of surfactant hydrocarbon tails was confirmed in electrokinetic experiments on nylon sols by Rendall *et al.*[19] For the alkyl sulphonates $C_{10}SO_3^-$, $C_{12}SO_3^-$ and $C_{16}SO_3^-$, the adsorption potentials were -6.4, -8.4 and $-12.9\,\mathbf{k}T$, respectively.

The shape of surfactant adsorption isotherms may be complex. An example is given in Fig. 2, showing the isotherms reported by Connor and Ottewill[35] for adsorption of a series of cationic quaternary alkyltrimethylammonium bromide surfactants with alkyl chain lengths C_8, C_{10}, C_{12} and C_{16} on an anionic polystyrene latex. The shape suggests that multistep reactions occur. At the first 'knee' of the C_{16} isotherm, the negative zeta potential was increased to zero. Hence the adsorption density of the hexadecyltrimethylammonium ion was taken to be equivalent to the number of surface charges on the latex. In this first region the cationic surfactants are counterions for the anionic surface charge. At higher surfactant concentrations, the zeta potential became positive; thus the slower increase in adsorption density was due to the electrostatic repulsion affecting adsorption of the cationic surfactants on the positive surface. Eventually, adsorption increases more steeply,

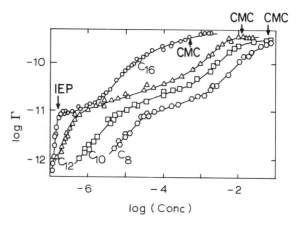

Fig. 2. Adsorption isotherms (Γ, mol cm^{-2}) of n-alkyltrimethylammonium ions on poly(styrene) latex at pH 8 at 10^{-3} M KBr as a function of concentration C for: $\bigcirc$, hexadecyltrimethylammonium, C_{16}; $\triangle$, dodecyltrimethylammonium; $\square$, decyltrimethylammonium; and ($\bigcirc$) octyltrimethylammonium ions. IEP indicates the isoelectric point for the C_{16} isotherm, and CMC indicates the critical micelle concentration (with permission [35]).

owing to the formation of surfactant aggregates at the latex/water interface. The C_8, C_{10} and C_{12} surfactants could be plotted on a single curve of a 'reduced' isotherm, i.e. Γ_i versus C_i/C_0, where C_0 is the CMC of the particular surfactant.

In the region of increased adsorption at concentrations below the CMC, the uptake of surfactant can be approximately described in terms of a GCSG isotherm,[32-34] e.g:

$$\Gamma_i = \frac{\Gamma_{max} \dfrac{C_i}{55 \cdot 5} \exp\left(-W_i/\mathbf{k}T\right)}{1 + \dfrac{C_i}{55 \cdot 5} \exp\left(-W_i/\mathbf{k}T\right)} \qquad (68)$$

where $W_i = z_i e \psi_i + m\Phi$, m is the number of methylene groups in the hydrophobe, and Φ is the hydrophobic energy per methylene group. This type of isotherm can be incorporated in the self-consistent GCSG models, where Γ_i contributes to the compact-layer charge σ_β, although no specific instances are reported.

2.5.2. Adsorption of Hydrolysable Metal Ions

The adsorption of alkali and alkaline-earth metal cations is generally thought to be described by the GCSG model with inclusion of some specific surface interactions. However, there are many divalent and higher valent metal ions (e.g. Cu(II), Pb(II), Co(II), Fe(III), Al(III) and Cr(III)) whose adsorption behaviour is enhanced and complicated by the formation of complex soluble or colloidal insoluble hydroxide species. A good example is the aqueous aluminium ion Al_{aq}^{3+}. Complex species may include mono or dihydroxo complexes (e.g. $Al(OH)^{2+}$ and $Al(OH)_2^{+}$[36,37]) as well as a range of polymeric species of various molecular weights (e.g. $Al_2(OH)_2^{2+}$,[38] $Al_8(OH)_{20}^{4+}$,[39] $Al_{13}O_4(OH)_{28}^{3+}$ [40] and $Al_{13}O_4(OH)_{24}(H_2O)_{12}^{7+}$ [41,42]), and various microcrystalline forms (e.g. $Al(OH)_3$, gibbsite, boehmite and bayerite). The amount of these species depends on the total concentration of Al(III), pH, ionic strength, temperature and the presence of complexing ligands.[43-45] A complete description of adsorption of such metal ions would require that the formation of complexes and solids be taken into account. The chemistry is complex and often slow to reach equilibrium. Many of the equilibrium constants for the reactions are not well known, and it has proved difficult to develop general adsorption models without invoking a number of simplifying assumptions.

However, some qualitative generalisations are possible.[46]

(1) For each hydrolysable metal, there is a characteristic, critical pH range, often $\Delta pH \sim 1$, over which the fraction of metal adsorbed increases from a low value to nearly unity. Adsorption behaviour depends partly on the substrate, but the major factor is the metal ion.

(2) Because adsorption increases with pH, the stoichiometry must involve release of protons or uptake of hydroxyl ions. This could be due to preferential adsorption of metal hydroxo species or to an exchange reaction where metal ions release surface-bound protons.

(3) Usually the increased surface activity occurs at a pH just lower than either the pK_a for the soluble complex or the pH where colloidal hydroxides are formed.

(4) The colloid stability and the zeta potential of a latex may change abruptly as adsorption occurs. As the uptake of metal complexes increases, cationic charge accumulates in the Stern layer σ_β, and this may neutralise the surface charge σ_0, causing the zeta potential to approach zero or even reverse its sign. Concurrently, the latex may coagulate or be redispersed with opposite zeta potential.

(5) As the pH is increased by the addition of base, the solubility product of hydroxides of adsorbed and dissolved complexes will be exceeded and a new colloidal phase will be formed on the latex surface. This may occur by condensation of adsorbed species or by heterocoagulation of latex and the newly formed metal hydroxide hydrosol. If the surface coverage is thick enough, the new surface on the latex reflects only the nature of the metal hydroxide. Because metal hydroxides are amphoteric, the coated latex may have a new isoelectric point.

An example of these features has been demonstrated by Matijevic and Bleier[47] who investigated the zeta potential and the colloid stability of PVC latex in aqueous Al^{3+} solutions. Their results are shown in Fig. 3. Similar results have been observed for rubber latex,[48] poly(styrene) latex[49] and poly(tetrafluoroethylene) latices.[50,51]

2.6. Adhesion or Heterocoagulation of Latices with Other Colloidal Particles

A subject related to adsorption in the electrical double layer is the mutual interaction of dissimilar particles in a mixture of colloids. An

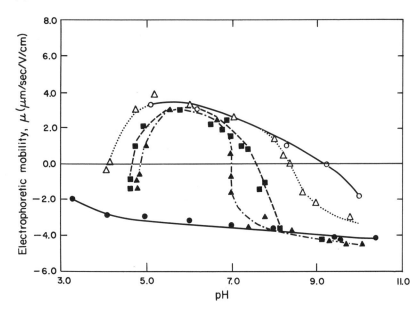

Fig. 3. Electrophoretic mobility of PVC latex (0·005 wt%) as a function of pH in the presence of $2·5 \times 10^{-5}$ mol dm^{-3} (▲), $1·0 \times 10^{-4}$ mol dm^{-3} (■), and $1·0 \times 10^{-3}$ mol dm^{-3} (△) Al(NO₃)₃ 24 h after mixing, compared with the mobilities of PVC latex (●) and aluminium hydroxide (○) alone (with permission[47]).

example is the interaction between anionic poly(styrene) particles and cationic aluminium hydroxide[45] when their dispersions are mixed at neutral pH. The oppositely charged particles rapidly coagulate each other. Other examples have been given for mixtures of different amphoteric latices.[25] As a result of such interactions, it is possible to control coagulation rates by adding one colloidal dispersion (e.g. a latex) to another colloid of different surface characteristics.[52–55]

3. EXPERIMENTAL METHODS

There are a number of standard electrochemical techniques for evaluating adsorption and desorption of ions at the latex/water interface. These include:

(1) Potentiometric acid–base titrations to evaluate the surface charge via protonation or deprotonation.

(2) Direct adsorption of ions (e.g. divalent cations) can also be investigated by use of ion-selective electrodes. Other analytical techniques such as isotopic labelling and exchange methods may also be used.

(3) Conductometric titrations to follow the change in ionic species as acids or bases are added to ionise the surface groups. The intersection of linear segmented conductance–volume curves gives the analytical endpoint.

(4) Measurement of the electrokinetic potential by a number of techniques, including microelectrophoresis and moving-boundary methods.

In addition, enthalpic titrations can be used to evaluate surface reactions, although few examples appear in the latex literature. All theoretical interpretations are made in terms of an adsorption density, i.e. Γ_i in mol m^{-2} or σ_0 in C m^{-2}. Hence supplementary data about the specific surface area or average particle size and particle-size distribution are essential and can be obtained. Fortunately, for narrow-size-distribution latices there are many methods based on, for example, electron microscopy, light scattering, gas adsorption and centrifugation.

3.1. Potentiometric Titration

The surface charge density of a latex with ionisable surface groups (e.g. —RCOOH or —N$^+$R$_2$H) can be defined as the net uptake of H$^+$ and OH$^-$ from a given solution composition, viz:

$$\sigma_0 = F(\Gamma_{H^+} - \Gamma_{OH^-}) = \frac{F}{A}(C_A - C_B - [H^+] + [OH^-]) \tag{69}$$

where F is the Faraday constant, A is the surface area in m^2 dm^{-3}, C_A and C_B are the concentrations of acid and base in the dispersion, and [H$^+$] and [OH$^-$] are the concentrations of H$^+$ and OH$^-$, obtained by pH measurements with a glass electrode and a double-junction reference electrode.

For the titration of an anionic latex with base, the following relation applies:

$$\sigma_0 = F[-COO^-]/A = F(-C_B - [H^+] - [OH^-])/A \tag{70}$$

Usually the latex is titrated with both acid and base to determine the reversibility of the surface reaction. Because the charge depends on

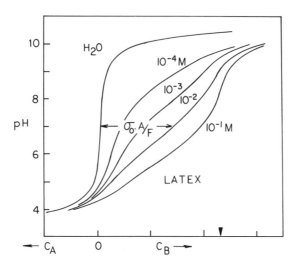

Fig. 4. Schematic illustration of a potentiometric titration of a latex at various ionic strengths compared with the titration curve for water. The difference between curves at constant pH is related to the surface charge density $\sigma_0(A/F)$.

electrolyte concentration, salt solutions must be added to 'maintain' constant ionic strength. The ionic strength can be checked by simultaneous conductivity measurements. When titrations are repeated for a number of ionic strengths, a family of curves is generated, as shown schematically in Fig. 4. For simple monofunctional latices that have no point of zero charge (PZC), it is essential to purify any added salts to remove traces of acids or bases that would affect the shape of the titration curves. The charge density can be obtained directly from Fig. 4 by taking, at a particular pH, the difference between the titration curve at a particular ionic strength and the background water (strong acid/strong base) titration curve. This gives the charge $+F(C_A - C_B - [H^+] + [OH^-])$ per area A (m^2). The titration data can then be analysed by the methods described earlier.

3.2. Conductometric Titration

In dilute solutions the specific conductivity κ (ohm^{-1} cm^{-1}) is evaluated from the individual ionic equivalent conductance λ_i (equiv^{-1} ohm^{-1} cm^2), the concentration C_i (mol dm^{-3}), and the magnitude of the charge $|z_i|$ for each component of the solution by:

$$\kappa = \sum_{\text{all ions}} \lambda_i C_i |z_i| / 1000 \quad \text{ohm}^{-1}\,\text{cm}^{-1} \tag{71}$$

Many ions involved in acid–base reactions on surfaces or in solutions have quite different values for λ_i, e.g:

$$\lambda_{H^+}^0 = 349\cdot8, \qquad \lambda_{OH^-}^0 = 198\cdot3, \qquad \lambda_{K^+}^0 = 73\cdot5, \qquad \lambda_{NO_3^-}^0 = 71\cdot4$$

at 25°C and infinite dilution. Thus conductometric titrations of strong acids and bases give curves with approximately linear segments as C_H decreases to the endpoint and then C_{OH} increases past the endpoint. These lines intersect at the endpoint. For weak acids or bases, where uncharged molecules may be present at significant concentrations, the shape of conductometric titration curves is more complex. The titration results of latices are also more complicated. As in simple solutions, the uptake and release of ions (e.g. H^+ or OH^-) from strong acid sites usually results in linear segmented curves, with the endpoint being a measure of surface sites. However, for this simple case and also for weak-acid surface sites, the charged particles themselves and the adsorbed ions in the electrical double layer can also contribute to the conductance. This is an area of active research, and the results are complicated.[56–58] However, if the volume fraction of latex is low (<0·05) and there is a low amount of added electrolyte, these extra contributions can probably be ignored if we are interested only in determining the site density and a qualitative estimate of surface acidity. As the ionic strength increases, the relative change in conductance becomes smaller, owing to the increased concentration of soluble charge carriers, e.g. K^+ and NO_3^-. In addition, the individual equivalent ionic conductance decreases, owing to the drag of the ionic atmosphere. Corrections to the limiting equivalent ionic conductance can be made by use of semi-empirical expressions,[59] viz:

$$\lambda_i = \lambda_i^0 - [B_1\lambda^0 + B_2]\sqrt{(I)}/\{1 + Ba\sqrt{(I)}\} \tag{72}$$

where for water at 25°C $B = 0\cdot3291 \times 10^8\,\text{cm}^{-1}$, $B_1 = 0\cdot23$, $B_2 = 30\cdot32$, and λ_i is the equivalent ionic conductance of ion i at total ionic strength I. At the hypothetical condition of infinite dilution, $\lambda_i = \lambda_i^0$. Figure 5 gives a schematic representation of conductometric titrations for a latex with a strong-acid surface group (e.g. alkyl sulphate) and a weak-acid group (e.g. alkyl carboxylate).

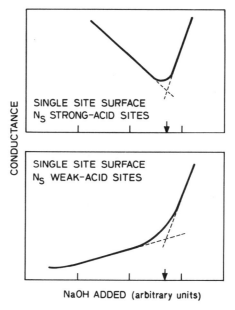

NaOH ADDED (arbitrary units)

Fig. 5. Schematic illustration of the conductometric titration of a monofunctional latex with strong-acid surface sites and weak-acid surface sites.

3.3. Electrokinetic Properties

The electrokinetic potential or zeta potential is a measure of the electrostatic potential at the hydrodynamic shear plane close to an interface and is influenced by the charge density inside the shear plane. Thus, when used in conjunction with an appropriate control measurement, zeta-potential measurements can yield useful information on the adsorption of ions at the colloid/water interface. The practical and theoretical aspects of determination of zeta potential have been summarised by Hunter.[7] For latices the determination of the zeta potential from the electrophoretic mobility or the particle velocity in an electric field is the most common technique. Hunter gives a full account of the effects of particle size and ionic strength on the conversion of mobilities to zeta potential. For particles larger than about $0 \cdot 1 \ \mu m$, which scatter sufficient light, microelectrophoresis with white light is satisfactory; however, for smaller particles, observation of light scattered by individual particles is too difficult unless laser illumination is used. For this reason, moving-boundary techniques have been used by some investigators working with small particles.[13] More recent techniques include laser Doppler spectroscopy of particles in an applied electric field.[7,60,61]

Zeta-potential data have been used as powerful information to complement direct adsorption measurements. Connor and Ottewill[35] determined the density of surface sites on an anionic latex by measuring the concentration of cationic surfactant required to neutralise the site charge, i.e. to reach the isoelectric point. From their adsorption isotherm, it was possible to obtain the adsorption density and hence adsorbed charge density for that particular concentration.

In another study of carboxylate latices, Ottewill and Shaw[22] estimated the surface acidity constants of the carboxylic acid. Their zeta-potential/pH data were similar in shape to a curve relating the ionisation fraction to the pH for a weak acid. The magnitude of the zeta potential, ζ, increased from near zero at low pH to a maximum value, ζ_{max}, in solutions of $pH > 7$.

Using an approximate diffuse-layer charge-potential relationship for small spherical particles of radius a, they obtained:

$$\sigma_d = D\varepsilon_0 \frac{\zeta}{a}(1 + \kappa a) \tag{73}$$

and

$$\sigma_{d_{max}} = D\varepsilon_0 \frac{\zeta_{max}}{a}(1 + \kappa a) \tag{74}$$

In the absence of Stern-layer charge, $\sigma_0 = -\sigma_d$, the ratio of ionised to un-ionised sites was taken as:

$$[S^-]/[HS] = \zeta/(\zeta_{max} - \zeta) \tag{75}$$

Using the logarithmic form of eqn (48), i.e:

$$pK_S = pH - \log[S^-]/[SH] + e\zeta/2 \cdot 303kT \tag{76}$$

and the condition that:

$$\zeta = \zeta_{max/2'} \quad \text{when} \quad pH = pH_{\zeta_{max/2}}$$

they obtained:

$$pK_S = pH_{\zeta_{max/2}} + \frac{e\zeta_{max}}{4 \cdot 606kT} \tag{77}$$

The pK_S values ranged from $4 \cdot 0 \pm 0 \cdot 2$ at 5×10^{-4} mol dm^{-3} NaCl to $4 \cdot 45 \pm 0 \cdot 1$ at 5×10^{-2} mol dm^{-3} NaCl solution. These values are about the same as pK_a values for soluble carboxylic acids. However, the effect of ionic strength on these pK_a estimates is opposite to the trend of estimates for the apparent pK_a from potentiometric titrations for increasing salt concentration.[14]

4. APPLICATION OF CONCEPTS TO ADSORPTION OF IONS ON LATICES

It has been seen that ideas and experimental techniques concerning adsorption at a number of interfacial systems can be extended to the latex/water interface. It remains to be demonstrated how models can describe or predict measurable properties such as the surface charge density or the adsorption density, the zeta potential, and the conductivity of dispersions. There are few studies in which all of these measurements are reported for one particular latex. One such study was the investigation of the surface charge and the zeta potential of poly(vinyl toluene) (PVT) latices by Bagchi et al.[13] This PVT latex had a low charge density ($\sim 0 \cdot 5$ $\mu C\,cm^{-2}$), owing to strongly acidic dodecyl sulphate functional groups. Under most pH conditions it was strongly ionised, and consequently there was little change in the magnitude of the surface charge or the zeta potential with pH. Both potentiometric and conductometric titration curves were typical of strong acid–base titrations. The surface charge and the zeta potential reported by Bagchi et al.[13] were adequately described by a diffuse double layer, ionised-surface model (see Section 2.4.1). Other studies of the adsorption of electrolytes and surfactants by use of conductance measurements combined with electrolyte replacement in a diafiltration apparatus have been described by Ahmed et al.[62]

The use of combined techniques is often necessary to investigate the electrical double layer structure of latices.[63,64] Generally, the interpretation of the behaviour of latices is simpler for particles with strong-acid surface groups. For latices with weak-acid surface groups, the ionisation tendency is more strongly dependent on the pH, the salt concentration and the total site density;[14,15] consequently, interpretation is more complicated.

An effort to demonstrate the use of EDL and solution-chemistry models was reported by James et al.[17] who analysed the data of Stone-Masui and Watillon[14,15] for latices stabilised by carboxylic acids. Some of their results are summarised below. These investigators[14]

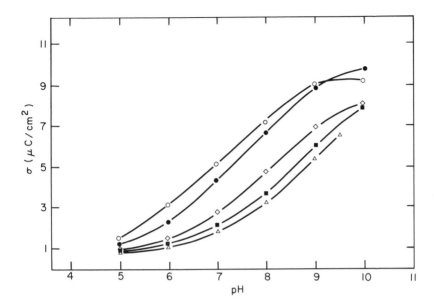

Fig. 6. Experimental surface charge density of a poly(styrene) latex prepared with potassium stearate as emulsifier as a function of pH and ionic strength controlled by addition of NaClO$_4$: $\triangle$, no added salt; $\blacksquare$, 10^{-4}; $\diamond$, 10^{-3}; $\bullet$, 10^{-2}; $\bigcirc$, 10^{-1} mol dm^{-3} NaClO$_4$ electrolyte concentration (with permission[14]).

reported acid–base titration data for a latex prepared with potassium stearate as the emulsifier. The shape of the titration curves was typical of weak-acid titration, as shown in Fig. 4. The surface charge density evaluated from these data varied with pH and ionic strength, as shown in Fig. 6. Because the pH for half neutralisation decreased with increased salt concentration, the apparent acidity of the functional groups increased with ionic strength.

From such data, James *et al.*[17] constructed a double extrapolation data set showing the variation of the acidity quotient Q_a with the fractional development of surface charge α and salt concentration $\sqrt{C}$, as outlined in eqn (65). This is reproduced in Fig. 7. Extrapolation of data at constant ionic strength (i.e. $\sqrt{C}$ constant) and at constant charge (i.e. α constant) yields at infinite dilution and zero charge the $pK_a^{int} = pQ$ ($\alpha = 0$, $\sqrt{C} = 0$) as the intercept on the ordinate. For this particular latex pK_a^{int} was 4·9, close to values for carboxylic acids in

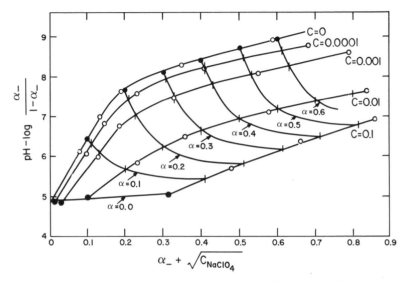

Fig. 7. Variation of the surface acidity quotient pQ_a as a function of surface charge and electrolyte concentration. Lines are contours of constant electrolyte concentration C or constant fractional ionisation, $\alpha = \sigma_0/(\sigma_{0_{max}} - \sigma_0)$.

homogeneous solution. A similar double extrapolation procedure was used to evaluate the site binding constant pK_M^{int} (eqn (66)). This time pQ_a was plotted as a function of both fractional charge α and concentration $\log_{10} C$. Extrapolation to zero charge and unit concentration gave $pK_M^{int} = 4.4$ as the intercept on the ordinate. These values for the intrinsic ionisation constants, the site density corresponding to $8\ \mu C\ cm^{-2}$ and the electrolyte concentrations were used in the site-binding, compact double layer model described by eqns (38)–(60) to obtain the calculated dependence of charge on pH shown in Fig. 8. These calculated curves generally give good agreement with the experimental data over a wide range of conditions. The potential of the diffuse double layer ψ_d is also obtained in this calculation. The values of ψ_d were in reasonable agreement with some data obtained earlier by Stone-Masui and Watillon[65] and by other workers,[22] although this is not a complete test of the self-consistency of the model.

The corresponding form of the acid–base titration of the latex is shown together with specific conductivity in Fig. 9 as a function of

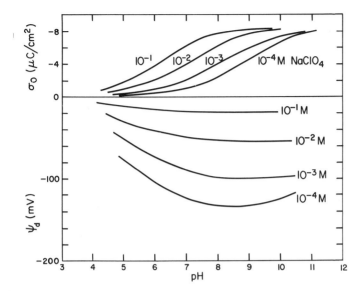

Fig. 8. Simulated charge σ_0 and diffuse layer potential ψ_d of a carboxylated latex as a function of pH and electrolyte concentration for EDL parameters $pK_a^{int} = 4 \cdot 9$, $pK_{Na}^{int} = 4 \cdot 4$ and $K_1 = 80 \ \mu F \ cm^{-2}$.

ionic strength. Characteristic titration curves of pH or conductivity against added base concentration are demonstrated; they indicate the apparent decrease of the surface acidity quotient pQ_a as the ionic strength increases.

The influence of increasing acid strength of surface group on the potentiometric and conductometric titrations is shown in Fig. 10 by substituting various values of pK_a^{int} and pK_M^{int} in the site-binding compact double layer model and then calculating the titration curves. The values for these parameters are given in Table 1. The results shown in Fig. 10 were calculated by assuming that no excess electrolyte was initially present. Thus the ionic strength does change during the titration. This change and the change in ion binding that contributes to the Stern-layer charge σ_β means that for these conditions there is no direct relationship between σ_0 and a diffuse-layer potential ψ_d or ζ.[17,21] Thus if counterion binding is significant, the method of determination of surface ionisation constants used by Ottewill and Shaw[22] by assuming a direct proportionality between σ_0 and ζ becomes less reliable.

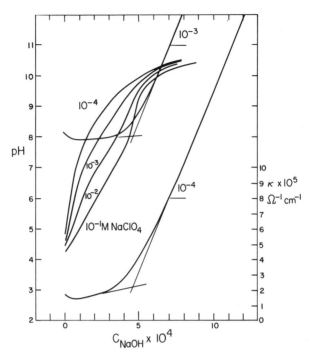

Fig. 9. Simulated potentiometric and conductometric titration of a carboxylated latex with NaOH in the presence of various concentrations of $NaClO_4$ for EDL parameters $pK_a^{int} = 4 \cdot 9$, $pK_{Na}^{int} = 4 \cdot 4$ and $K_1 = 80 \ \mu F \ cm^{-2}$.

TABLE 1

Test Values for Surface Ionisation Constants and Compact-Layer Capacitances for an Aqueous Latex

Latex	pK_a^{int}	pK_M^{int}	K_1 ($\mu F \, cm^{-2}$)	K_2 ($\mu F \, cm^{-2}$)
a	$4 \cdot 9$	$4 \cdot 4$	80	20
b	$3 \cdot 0$	$3 \cdot 0$		
c	$2 \cdot 0$	$1 \cdot 5$		
d	$1 \cdot 0$	$1 \cdot 0$		
e	$0 \cdot 5$	$0 \cdot 0$		
f	$-1 \cdot 0$	$-1 \cdot 0$		

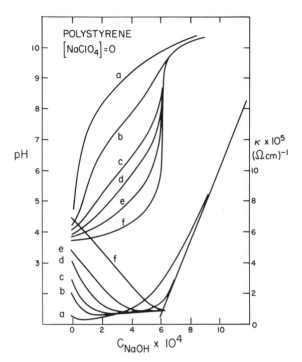

Fig. 10. Effect of increasingly acidic surface functional groups on the simulated potentiometric and conductometric titration of a poly(styrene) latex. The ionisation constants are given in Table 1; acidity increases from curve a to curve f. Curve a corresponds to carboxylate latex (Figs 6 and 9), and curve f corresponds to the behaviour of sulphonate latices.

5. CONCLUDING REMARKS ON SIMPLE ION ADSORPTION

Surface site ionisation models appear to account for a number of the characteristic properties of latex/water interfaces in a reasonably self-consistent manner. For latices with low charge density (e.g. $\sim 0.5 \, \mu C \, cm^{-2}$), such models using a diffuse double layer seem to be adequate. On the other hand, when the site density is higher (e.g. $\sim 5 \, \mu C \, cm^{-2}$), the Gouy–Chapman–Stern–Grahame model of a compact and a diffuse double layer charge appears to be more appropriate. Estimates of the conductivity titration curves can also be made by considering only those ions in the equilibrium bulk solution. This is probably sufficient to estimate site density endpoints and acidity of surface groups. Increased performance may be obtained in the future as theories accounting for the contribution of EDL ions and latex particles to the conductivity become simpler to apply. These models now incorporate many of the features and parameters discussed earlier, for example reaction tendencies (e.g. K_a^{int}), surface sites and limiting site densities, the effects of the electrical double layer on adsorption, and a stoichiometry for the interfacial reactions. This discussion has focussed on adsorption at a surface. Under some conditions, absorption of ions or surfactants may occur and prove difficult to distinguish from adsorption; if this is the case, the term sorption is preferable. Of course, if absorption does occur, then the models described here must be modified to describe the distribution of ions and molecules inside the latex particles.

REFERENCES

1. Langmuir, I., *Chem. Rev.*, **13**, 147 (1933).
2. Adamson, A. W., *Physical chemistry of surfaces*, 2nd edn, Wiley, New York (1967).
3. Parks, G. A., In *Chemical oceanography*, Vol. 1, 2nd edn, J. P. Riley and G. Skirrow (Eds), Academic Press, New York (1975), p. 241.
4. Kitchener, J. A., *J. Photogr. Sci.*, **13**, 152 (1965).
5. Gouy, G., *J. Phys. Chem.*, **9**, 457 (1910).
6. Chapman, D. L., *Philos. Mag.*, **25**, 475 (1913).
7. Hunter, R. J., *Zeta potential in colloid science, principles and applications*, Academic Press, London (1981).
8. Hiemenz, P. C., *Principles of colloid and surface chemistry*, Marcel Dekker, New York (1977).
9. Grahame, D. C., *Chem. Rev.*, **44**, 441 (1947).
10. Loeb, A. L., Overbeek, J. Th. G. and Wiersema, P. H., *The electrical double layer around a spherical colloid particle*, M.I.T. Press, Cambridge (1961).
11. White, L. R., *J. Chem. Soc. Faraday Trans. II*, **73**, 577 (1977).
12. Ohshima, H., Healy, T. W. and White, L. R., *J. Colloid Interface Sci.*, **90**, 17 (1982).
13. Bagchi, P., Gray, B. V. and Birnbaum, S. M., *J. Colloid Interface Sci.*, **69**, 502 (1979).
14. Stone-Masui, J. and Watillon, A., *J. Colloid Interface Sci.*, **52**, 479 (1975).
15. Watillon, A. and Stone-Masui, J., *J. Electroanal. Chem.*, **37**, 143 (1972).
16. Stern, O., *Z. Electrochem.*, **30**, 508 (1924).
17. James, R. O., Davis, J. A. and Leckie, J. O., *J. Colloid Interface Sci.*, **65**, 331 (1978).

18. Rendall, H. M. and Smith, A. L., *J. Chem. Soc. Faraday Trans.* I, **74**, 1179 (1978).
19. Rendall, H. M., Smith, A. L. and Williams, L. A., *J. Chem. Soc. Faraday Trans.* I, **75**, 669 (1979).
20. Healy, T. W. and White, L. R., *Adv. Colloid Interface Sci.*, **9**, 303 (1978).
21. James, R. O. and Parks, G. A., Chapter 2 in *Surface and colloid science*, Vol. 12, E. Matijevic (Ed.), Plenum Press, New York (1982), p. 119.
22. Ottewill, R. H. and Shaw, J. N., *Kolloid Z. Z. Polym.*, **218**, 34 (1970).
23. Goddard, E. D., *Adv. Colloid Interface Sci.*, **4**, 45 (1974).
24. Homola, A. and James, R. O., *J. Colloid Interface Sci.*, **59**, 123 (1977).
25. James, R. O., Homola, A. H. and Healy, T. W., *J. Chem. Soc. Faraday Trans.* I, **73**, 1436 (1977).
26. Harding, I. H. and Healy, T. W., *J. Colloid Interface Sci.*, **89**, 185 (1982).
27. Huang, C. P. and Stumm, W., *J. Colloid Interface Sci.*, **43**, 409 (1973).
28. Lyklema, J., *J. Colloid Interface Sci.*, **58**, 242 (1977).
29. Westall, J., *Adv. Chem. Ser.*, **189**, 33 (1980).
30. Westall, J., Zachary, J. L. and Morel, F. M. M., Technical Note No. 18, Water Qual. Lab., Dept of Civil Eng., Massachusetts Institute of Technology (1976).
31. Smit, W. and Holten, C. L. M., *J. Colloid Interface Sci.*, **78**, 1 (1980).
32. Healy, T. W., *J. Macromol. Sci. Chem.*, **8**, 603 (1974).
33. Somasundaran, P., Healy, T. W. and Fuerstenau, D. W., *J. Phys. Chem.*, **68**, 3562 (1964).
34. Somasundaran, P., Healy, T. W. and Fuerstenau, D. W., *J. Colloid Interface Sci.*, **22**, 599 (1966).
35. Connor, P. and Ottewill, R. H., *J. Colloid Interface Sci.*, **37**, 642 (1971).
36. Smith, R. W., *Adv. Chem. Ser.*, **106**, 251 (1971).
37. May, H. M., Helmke, P. A. and Jackson, M. L., *Geochim. Cosmochim. Acta*, **43**, 861 (1979).
38. Turner, R. C., *Can. J. Chem.*, **53**, 2811 (1975).
39. Hayden, P. L. and Rubin, A. J., in *Aqueous environmental chemistry of metals*, A. J. Rubin (Ed.), Ann Arbor Science, Ann Arbor (1974), p. 317.
40. Bottero, J. Y., Tchoubar, D., Cases, J. M. and Fiessinger, F., *J. Phys. Chem.*, **86**, 3667 (1982).
41. Bottero, J. Y., Cases, J. M., Fiessinger, F. and Poirer, J. E., *J. Phys. Chem.*, **84**, 2933 (1980).
42. Akitt, J. W., Greenwood, N. N., Khandewal, B. L. and Lester, G. D., *J. Chem. Soc. Dalton Trans.*, 604 (1972).
43. Brace, R. and Matijevic, E., *J. Inorg. Nucl. Chem.*, **35**, 3691 (1973).
44. Scott, W. B. and Matijevic, E., *J. Colloid Interface Sci.*, **66**, 447 (1978).
45. Lo, C-C., Meites, L. and Matijevic, E., *Anal. Chim. Acta*, **139**, 197 (1982).
46. James, R. O. and Healy, T. W., *J. Colloid Interface Sci.*, **40**, 42, 53, 65 (1972).
47. Matijevic, E. and Bleier, A., *Croat. Chem. Acta*, **50**, 93 (1977).
48. Matijevic, E. and Force, C. G., *Kolloid Z. Z. Polym.*, **225**, 33 (1968).
49. Ottewill, R. H. and Shaw, J. N., *J. Colloid Interface Sci.*, **26**, 110 (1968).
50. Ottewill, R. H. and Rance, D. G., *Croat. Chem. Acta*, **52**, 1 (1979).

51. Kratohvil, S. and Matijevic, E., *J. Colloid Interface Sci.*, **57,** 104 (1976).
52. Hansen, F. H. and Matijevic, E., *J. Chem. Soc. Faraday Trans. I*, **76,** 1240 (1980).
53. Bleier, A. and Matijevic, E., *J. Colloid Interface Sci.*, **55,** 510 (1976).
54. Bleier, A. and Matijevic, E., *J. Chem. Soc. Faraday Trans., I*, **74,** 1346 (1978).
55. Sasaki, H., Matijevic, E. and Barouch, E., *J. Colloid Interface Sci.*, **76,** 319 (1980).
56. O'Brien, R. W., *J. Colloid Interface Sci.*, **81,** 234 (1981).
57. Delacey, E. H. B. and White, L. R., *J. Chem. Soc. Faraday Trans. I*, **77,** 2007 (1981).
58. Saville, D. A., *J. Colloid Interface Sci.*, **91,** 34 (1983).
59. Robinson, R. A. and Stokes, R. H., *Electrolyte solutions*, 2nd edn, Butterworths, London (1965), p. 143.
60. Uzgiris, E. E., *Adv. Colloid Interface Sci.*, **14,** 75 (1981).
61. Ware, B. R., *Adv. Colloid Interface Sci.*, **4,** 1 (1974).
62. Ahmed, S. M., El-Aasser, M. S., Pauli, G. H., Poehlein, G. W. and Vanderhoff, J. W., *J. Colloid Interface Sci.*, **73,** 388 (1982).
63. Baran, A. A., Dudkina, L. M., Soboleva, N. M. and Checik, O. S., *Kolloidn. Zh.*, **43,** 211 (1981).
64. Laaksonen, J., LeBell, J. C. and Stenius, P., *J. Electroanal. Chem.*, **64,** 207 (1975).
65. Stone-Masui, J. and Watillon, A., *J. Colloid Interface Sci.*, **28,** 187 (1968).

Chapter 4

Adsorption from Solution—
Part II: Macromolecular Adsorbates

THARWAT F. TADROS

Plant Protection Division, ICI PLC, Bracknell, Berkshire, UK

1. INTRODUCTION

The use of both natural and synthetic polymers to control the stability behaviour of colloidal dispersions and suspensions is of considerable technological importance. Polymers find use as stabilisers in pharmaceutical, cosmetic, dyestuff and pesticide suspensions but nowhere is their use more widespread than in the preparation and application of polymer dispersions. They provide a particularly robust form of stabilisation which is useful at high disperse phase volume fractions and high electrolyte concentrations, as well as under extreme conditions of temperature, pressure and flow. The way in which the macromolecule adsorbs on the surface, the total amount that can be adsorbed, and the degree of extension and configuration of the chain near the interface are all key factors that determine their effectiveness in each particular application. It is the ability of polymers to adopt many conformations both in bulk solution and at an interface which dominates their technological adaptability, and in fact provides the key to a theoretical understanding of their activity. Thus in contrast to theories which describe the adsorption of low molecular weight species discussed in Chapter 3, a successful theory of polymer adsorption has to be intimately concerned with the details of intramolecular arrangements near the interface and the way in which these are perturbed by environmental factors.

The literature on polymer adsorption is vast with regard to both theoretical and experimental studies. Several monographs and review

105

articles are now available,[1-10] some of them fairly recent. For this reason, the present section on this subject will be brief, only highlighting those features which have already been adequately reviewed and only dealing in some detail with those more recent theories and experiments which have not been covered in the previous reviews.

2. GENERAL FEATURES AND PLAN

The process of polymer adsorption is fairly complicated. In addition to the usual adsorption considerations such as polymer–surface, polymer–solvent and solvent–surface interactions, one of the principal problems to be resolved is the conformation of the polymer molecule on the surface. This was recognised in 1951 by Jenckel and Rumbach,[11] who found that the amount of polymer adsorbed per unit area of the surface would correspond to a layer several molecules thick if all the segments of the chains are attached. They suggested a model in which each polymer molecule is attached in sequences separated by loops which extend into solution. In other words, not all the segments of a macromolecule are in contact with the surface. Sequences of segments which are in direct contact with the surface are termed 'trains'; those in between and extending into solution are termed 'loops'; the free ends of the molecule, also extending into solution, are termed 'tails'. However, this picture (schematically shown in Fig. 1(a)) of a random sequence of tails, trains and loops only applies to the case of a homopolymer molecule on a plane surface. Various other configurations may be distinguished depending on the structure of the polymer and the nature of the groups involved. For example, if the chain contains groups with a preferential affinity for the surface, the configuration is no longer random since these groups will form the adsorption points. This is illustrated in Fig. 1(b), whereby the preferentially adsorbed groups are in relatively short blocks. Clearly if all groups have a strong affinity for the surface, the polymer chain will adopt a flat configuration (Fig. 1(c)) with most (or all) of the surface sites occupied. This situation is seldom realised in practice since both kinetic and entropic reasons dictate that a considerable fraction of the segments will still be in loops or tails.

On the other hand, if the polymer molecule is formed from two blocks with large differences in their affinity to the surface, i.e. of the AB block type (with B having the larger affinity), the configuration

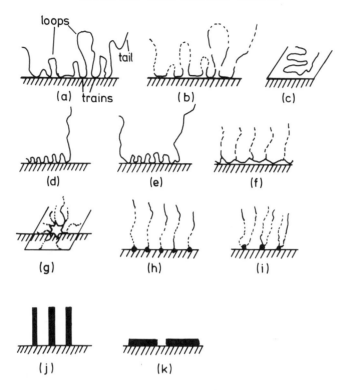

Fig. 1. Various conformations of macromolecules adsorbed on a plane surface. (a) Random conformation of loops–tails–trains (homopolymer); (b) preferential adsorption of short 'blocks'; (c) chain lying totally on the surface; (d) A block with loop–train configuration for B and one long tail for A; (e) ABA block as with (d); (f) BA_n graft with backbone (B) forming small loops leaving tails of A ('teeth'); (g) 'star' shaped molecule; (h) single point attachment at chain end; (i) single point attachment at middle of chain; (j) 'rod' shaped molecule lying vertical; (k) rod shaped molecule lying horizontal.

shown in Fig. 1(d) may result, whereby the B group adopts a loop–train conformation (with small loops) leaving all the A segments dangling in solutions as one long tail. A similar configuration is obtained with an ABA block (Fig. 1(e)) except in this case two long tails are present. A variance of this configuration is that obtained with a graft copolymer of the BA_n type (sometimes referred to as a 'comb' type structure) whereby now the B backbone adsorbs on the surface (either flat or with a 'loopy' conformation) leaving a number of dangling tails (sometimes referred to as 'teeth') in solution (Fig. 1(f)). A similar structure is obtained with a 'star' shaped polymer whereby a number of tails originate from around a central (Fig. 1(g)) block. In certain cases, the whole configuration may be totally formed by tails (single point attachment) (Fig. 1(h)) as, for example, is the case with a polymer molecule terminating with a functional group that is chemically bonded to the surface. A slight variation of this (Fig. 1(i)) is where the functional group is in the middle of the chain, whereby now two tails represent the configuration. Other types of configuration are obtained with 'rod' shaped molecules whereby the molecules are either vertically (Fig. 1(j)) or horizontally (Fig. 1(k)) attached.

It is clear from the above description of polymer conformations that for a full description of polymer adsorption it is necessary to answer the following questions. (1) What is the amount of polymer that is adsorbed per unit area of the surface (Γ)? (2) What groups are in direct contact with the surface and what fraction (p) of the total number of segments are these and how strongly are they adsorbed? (3) What is the distribution of polymer segments ($\rho(z)$) from the surface towards the bulk solution? How far do these segments extend in solution and what is the 'thickness' (δ) of the adsorbed layer? It is important to know how the above parameters change with polymer concentration/coverage, the structure of the polymer and its molecular weight, on the one hand, and any change in its environment (e.g. temperature, solvency, etc.) on the other.

In the next section, the theories of polymer adsorption will be mentioned including a brief account of the most recent and perhaps most complete treatment so far. The implications of such theories and their predictions will be described. A comparison of theoretical predictions with experimental results will be given. This is then followed by a description of the experimental quantities needed for describing polymer adsorption and how these can be obtained. As mentioned in the Introduction, this part on polymer adsorption is by no means comprehensive (due to lack of space) and will only serve as a guideline for the reader. For more comprehensive information on the subject, the reader should refer to the review articles and monographs[1-10] published and the references quoted therein. It should be mentioned that this subject is still growing and, therefore, it is highly likely that further useful papers will appear on both the theory and the experiments.

3. THEORIES OF POLYMER ADSORPTION

There are basically two main approaches to the theoretical treatment of polymer adsorption. The first, referred to as the random-walk approach, is based on the same concepts used for deducing the conformation of macromolecules in solution.[12] Basically the solution is represented by a three-dimensional lattice and the surface by a two-dimensional lattice. The polymer is represented by a realisation of a random walk on the lattice. The probability of performing steps in different directions is considered to be the same except at the interface which acts as a reflecting barrier. The second approach to the theoretical treatment of polymer adsorption, referred to as the statistical mechanical approach, treats the polymer configuration as being made up of three types of structure: trains, i.e. sequences of segments all of which are in the interface; loops, i.e. sequences of segments that have only their first and last segments in the interface; and tails, i.e. sequence of segments that start at the interface but never return. The units in these structures are considered to be in two different energy states, those close to the surface (i.e. trains) are adsorbed with an internal partition function determined by the short-range forces between the segments and the surface, whereas the units on loops and tails are considered to have an internal partition function equivalent to that of segments in bulk solution. By equating the chemical potential of the macromolecule in the adsorbed state and in bulk solution, the adsorption isotherm may be determined.

The earliest theories of polymer adsorption were those of Simha and co-workers,[13,14] Silberberg,[15] DiMarzio,[16] DiMarzio and McCrackin,[17] Hoeve et al.,[18] Rubin,[19] Roe[20] and Motomura and Matuura.[21] All these 'older' theories treat the case of an isolated chain on a surface, thus neglecting the interaction between the segments. Although some of these theories provide a suitable starting point for more realistic models, they have little relevance for practical systems. Even in very dilute solutions, the segment volume fraction near the interface is usually of the order of 0·5 so that the interaction between the segments plays a dominant role. Later theories which took this interaction into account were those of Silberberg[22] and Hoeve.[23] Unfortunately, such theories make assumptions about the segment density distribution which are not completely warranted. For example, Silberberg[22] used a step function whereas Hoeve[23] used an exponential function for the segment concentration profile in the loops. Moreover, in these treatments a surface phase with only adsorbed molecules[22] was assumed and tails were neglected.[23] Roe[24] introduced a more comprehensive theory of polymer adsorption, based on a simplifying assumption that each of the segments of a chain gives the same contribution to the segment density at any distance from the surface. In this way he was able to derive the segment density profile near the surface, although he did not calculate the loop, train and tail size distributions. The most recent theory is that of Scheutjens and Fleer[25–28] which takes into account all possible chain conformations with and without tails. Only this most recent theory will be summarised below, since all other treatments have been adequately reviewed.[1–10]

The basic procedure in Scheutjens and Fleer theory[25–28] is to describe all chain conformations as step-weighted random walks on a quasi-crystalline lattice which extends in parallel layers away from the surface. The partition function is written in terms of numbers of chain conformations (rather than in terms of individual segments used by Roe[24]), which are treated as connected sequences of segments. The interaction between segment and solvent molecules is taken into account using the Bragg–Williams approximation of random mixing within each layer parallel to the surface (i.e. in a manner similar to the well-known Flory–Huggins theory of concentrated polymer solutions). Each step in the random walk is assigned a weighting factor p_i which is considered to consist of three contributions: the adsorption energy, the configurational entropy of the solvent and the segment–solvent interaction energy.

For a chain of r segments and $(r-1)$ bonds, the conformation probability $P(c)$ is given by the following expression:

$$P(c) = \lambda_0^q \lambda_1^{r-1-q} p_1^{r_1} p_2^{r_2} \ldots p_i^{r_i} \ldots p_M^{r_M} \tag{1}$$

where $r_1, r_2, \ldots r_i \ldots r_M$ are the number of segments in layers $1, 2, \ldots i, \ldots M$ (where $i = 1$ is the layer adjacent to the surface and $i = M$ is the bulk solution) and $p_1, p_2, \ldots p_i, \ldots p_M$ are their segmental weighting factors. q is the number of bonds parallel to the surface and λ_0 and λ_1 are lattice parameters that give the fraction of neighbours within one layer λ_0 and in adjacent layers λ_1, with $\lambda_0 + 2\lambda_1 = 1$. For a simple cubic lattice $\lambda_0 = 4/6$ and $\lambda_1 = 1/6$ whereas for a hexagonal lattice $\lambda_0 = 6/12$ and $\lambda_1 = 3/12$.

The probability of any conformation can be calculated from a knowledge of the segmental weighting factors. In turn, the relative contribution from each chain conformation to the segment concentra-

Trimers

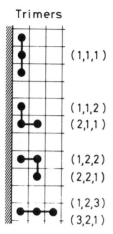

(1,1,1)

(1,1,2)
(2,1,1)

(1,2,2)
(2,2,1)

(1,2,3)   **Fig. 2.** Possible chain conformations for an
(3,2,1)   adsorbed trimer.[28]

tion in each layer can be used to calculate the volume fraction in each layer thus enabling one to calculate the segment concentration profile and hence the adsorbed amount. To illustrate how this can be done, let us consider a very simple case of a chain consisting of only three segments. An adsorbed trimer will have one of the four possible conformations shown in Fig. 2 in which a conformation is indicated by listing sequentially the number of the layer, where each of the three successive chain segments of the trimer finds itself. The conformational probabilities, according to eqn (1), are given by:

$$P(1, 1, 1) = \lambda_0^2 p_1^3; \qquad P(1, 1, 2) = P(2, 1, 1) = \lambda_0 \lambda_1 p_1^2 p_2 \qquad (2a)$$

$$P(1, 2, 2) = P(2, 2, 1) = \lambda_0 \lambda_1 p_1 p_2^2; \qquad P(1, 2, 3) = P(3, 2, 1) = \lambda_1^2 p_1 p_2 p_3 \qquad (2b)$$

The volume fraction of segments in each layer ϕ_i relative to that in bulk solution ϕ_* is obtained from the proper weighting of each of the above conformations in that layer. Therefore conformation (1, 1, 1) has all its segments in the first layer, thus contributing to ϕ_1 with a weight $P(1, 1, 1)$. On the other hand, conformations (1, 1, 2) and (2, 1, 1) have only two segments in layer 1, thus contributing to ϕ_1 with a weight $\frac{2}{3}P(1, 1, 2)$ each, i.e. $\frac{4}{3}P(1, 1, 2)$. Conformations (1, 2, 2) and (2, 2, 1), and (1, 2, 3) and (3, 2, 1) have only one segment in layer 1, thus contributing with a weight $\frac{1}{3}P(1, 2, 2)$, $\frac{1}{3}P(2, 2, 1)$, $\frac{1}{3}P(1, 2, 3)$,

$\frac{1}{3}P(3, 2, 1)$, respectively. Thus:

$$\phi_1/\phi_* = P(1, 1, 1) + \tfrac{4}{3}P(1, 1, 2) + \tfrac{2}{3}P(1, 2, 3) + \tfrac{2}{3}P(1, 2, 2) \qquad (3a)$$

The corresponding volume fractions in layers 2 and 3 are given by:

$$\phi_2/\phi_* = \tfrac{2}{3}P(1, 1, 2) + \tfrac{4}{3}P(1, 2, 2) + \tfrac{2}{3}P(1, 2, 3) \qquad (3b)$$

$$\phi_3/\phi_* = \tfrac{2}{3}P(1, 2, 3) \qquad (3c)$$

where the chain conformation probabilities $P(1, 1, 1)$, etc. are given in eqn (2).

As mentioned previously, the segmental weighting factor p_i is made up from three contributions: (1) an entropy term due to the transfer of solvent molecules from a layer i to bulk solution; (2) a contact energy term accounting for the different surrounding of a segment (and solvent) in the bulk solution and in layer i; and (3) an adsorption energy term due to the interaction of a segment with the surface. Clearly, the latter term is only present in the first layer. The configurational entropy loss $\Delta S°$ due to the transfer of a solvent molecule from a layer i with solvent volume fraction $\phi_i°$ to bulk solution with a solvent volume fraction $\phi_*°$ (note that $\phi_*° > \phi_i°$) is given by $\mathbf{k} \ln \phi_*°/\phi_i°$ per solvent molecule. This entropy term can be written as a Boltzmann factor $\exp(-\Delta S°/\mathbf{k}) = \phi_i°/\phi_*°$ in the weighting factor p_i.

The transfer of a segment from bulk solution to layer i is accompanied by an energy change (in $\mathbf{k}T$ units) of $\chi(\phi_i° - \phi_*°)$, where χ is the Flory–Huggins segment–solvent interaction parameter. This expression is only valid for the case of a homogeneous region around a site in layer i. Since an adsorbed layer is not homogeneous in the direction perpendicular to the surface, a correction is necessary to account for the fact that only a fraction λ_0 of the contacts are within layer i, and a fraction λ_1 with each of the neighbouring layers where the concentrations $\phi_{i-1}°$ and $\phi_{i+1}°$ are different. Therefore, the mixing energy per segment becomes $\chi(\langle\phi_i°\rangle - \phi_*°)$ where $\langle\phi_i°\rangle$ is the weighted average of the volume fractions in layers $(i-1)$, i and $(i+1)$, i.e.:

$$\langle\phi_i°\rangle = \lambda_1\phi_{i-1}° + \lambda_0\phi_i° + \lambda_1\phi_{i+1}° \qquad (4)$$

Similarly the transfer of a solvent molecule from layer i to bulk solution is accompanied by an energy change $-\chi(\langle\phi_i\rangle - \phi_*)$. Since $\phi_i + \phi_i° = 1$ for any layer i, $\langle\phi_i\rangle + \langle\phi_i°\rangle = 1$ for all layers except the first. Hence, the sum of the two mixing energies is equal to $2\chi(\langle\phi_i°\rangle - \phi_*°)$ corresponding to a Boltzmann factor $\exp\{-2\chi(\langle\phi_i°\rangle - \phi_*°)\}$ in p_i $(i > 1)$.

For the layer adjacent to the surface $\langle\phi_1^\circ\rangle + \phi_1 = 1 - \lambda_1$ according to eqn (4).

The adsorption energy in the first layer gives rise to a Boltzmann factor $\exp \chi_s$ in the weighting factor, where χ_s is the difference in adsorption energy (in kT units) for a segment and for a solvent molecule.

Thus, the weighting factors p_1 and p_i in the first and subsequent layers are given by the expressions:

$$p_1 = (\phi_1^\circ/\phi_*^\circ) \exp\{-2\chi(\langle\phi_i^\circ\rangle - \phi_*^\circ) \exp(\chi_s - \lambda_1\chi)\} \tag{5}$$

$$p_i = (\phi_i^\circ/\phi_*^\circ) \exp\{-2\chi(\langle\phi_i^\circ\rangle - \phi_*^\circ)\} \quad i \geq 2 \tag{6}$$

The term $\lambda_1\chi$ in eqn (5) comes in automatically (due to $\langle\phi_1^\circ\rangle + \phi_1 = 1 - \lambda_1$) and accounts for the fact that a segment in one of the surface layers is deprived of a fraction λ_1 of its contacts with 'solution' sites, these having been replaced by contacts with the surface.

From the weighting factors (eqns (5) and (6)), the statistical weight of any chain conformation can be calculated for a given segment concentration profile $\phi_1^\circ, \phi_2^\circ, \phi_3^\circ \ldots \phi_*^\circ$. Using a matrix formulation first introduced by Rubin[19] and DiMarzio and Rubin,[29] these statistical weights can be used to calculate the overall segment concentration profile $\phi_1, \phi_2, \phi_3 \ldots \phi_*$ in the adsorbed layer. Moreover Scheutjens and Fleer[25-28] have also calculated the bound fraction p and direct surface coverage θ_1, the distribution of trains, loops and tails as well as the root mean square adsorbed layer thickness. In the next section, the theoretical predictions and comparison with experimental results are briefly discussed.

4. THEORETICAL PREDICTIONS AND COMPARISON WITH EXPERIMENTAL RESULTS

A complete theory on polymer adsorption should provide information on three main points: (1) the relationship between the adsorbed amount and polymer concentration in bulk solution; (2) the bound fraction of segments, i.e. those in trains; and (3) the structure of the adsorbed layer, i.e. the distribution of individual segments from the surface towards bulk solution and the fraction of segments in loops and tails and their sizes. It is necessary to show how the above variables change with surface coverage, molecular weight of the polymer, poly-dispersity of the chain and solvency of the medium. A brief account of the predictions of Scheutjens and Fleer's theory in relation to these major variables is given below. This is then followed by a comparison with experimental results, where reliable measurements are available. It should be mentioned, however, that quantitative experimental verification of the theoretical treatment has, as yet, not been achieved due to the difficulty of measuring accurately all the key parameters simultaneously. As we will see below, significant progress has been made in recent years, but still the theory is far ahead of experiments.

4.1. Theoretical Predictions

4.1.1. Adsorption Isotherms
A useful presentation of adsorption isotherms was that given by Scheutjens and Fleer[28] in which the adsorbed amount, expressed as the total surface coverage θ given in equivalent monolayers, was plotted as a function of polymer volume fraction ϕ_* in bulk solution. The latter was varied between 0 and 1, thus covering the whole concentration range and the isotherms were given as a function of the number of segments in the chain r, which was varied from $r = 1$ to $r = 5000$, thus covering a very wide range from monomers and oligomers to fairly high molecular weight polymers with 5000 segments. The influence of solvency was also included in this presentation by using two values of χ, namely $\chi = 0$ (athermal solvent) and $\chi = 0.5$ (theta solvent). The segment–surface adsorption energy χ_s was taken to be $1\,\mathbf{k}T$ unit. The results of these calculations are shown in Fig. 3. A number of interesting features can be deduced from these isotherms. Oligomers adsorb according to a Langmuir-type isotherm with a low affinity, as expected for this rather low adsorption energy. The longer the chain (i.e. the higher the molecular weight), the more pronounced the high affinity character becomes and the more θ increases with ϕ_*. At $\phi_* = 1$ (i.e. bulk polymer) and $r \geq 5$, θ is proportional to the square root of chain length. This is to be expected from the Gaussian distribution of the chains in bulk polymer (chain dimension $\propto r^{\frac{1}{2}}$) where excluded volume effects are no longer important. The effect of solvent quality on adsorption is important in relatively dilute solutions, whereby the segment concentration in the adsorbed layer is appreciably different from that in bulk solution. Under these conditions, the adsorbed amount increases with decrease of solvency and this increase is significant for all 'polymers' containing ten or more segments.

In practice, adsorption isotherms are usually obtained over a limited

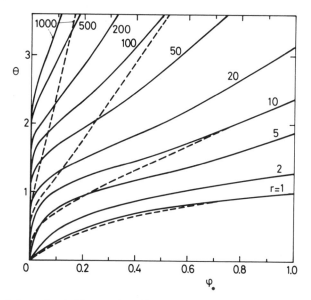

Fig. 3. Adsorption isotherms for oligomers and polymers for the whole range of solution volume fraction.[28] ——, $\chi = 0\cdot5$; − − −, $\chi = 0$, $\chi_s = 1$; hexagonal lattice ($\lambda_0 = 0\cdot5$).

range of polymer volume fraction in bulk solution (of the order of 10^{-3} and lower) and therefore a limited representation is shown in Fig. 4. Here the surface coverage in equivalent monolayers (θ) is plotted as a function of ϕ_* for values between 0 and 10^{-3} for a wide range of chain lengths under both athermal and theta solvency conditions. For $1\,\mathbf{k}T$ adsorption energy θ is very small when $r = 1$; the isotherm does not deviate much from a straight line when $r = 10$ (beyond the Henry region) and for $r > 20$ high-affinity isotherms are obtained. The adsorption isotherms for chains with $r = 100$ and above are typical of those observed experimentally for most polymers which are not too poly-disperse, i.e. showing a steep initial rise followed by a nearly horizontal pseudoplateau (which only increases a few per cent per decade of ϕ_*). In these dilute solutions the effect of solvency is most clearly seen, with poor solvents giving the highest adsorbed amounts. In good solvents (see dashed lines in Fig. 4) θ is much smaller and levels off for long chains to attain an adsorption plateau which is essentially independent of molecular weight. It should be noted that in most real situations the

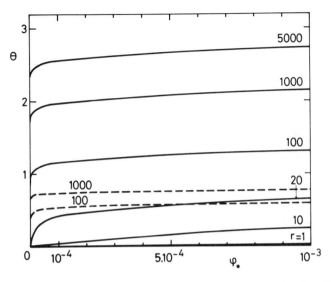

Fig. 4. Adsorption isotherms for oligomers and polymers in the dilute region.[28] ——, $\chi = 0.5$; ---, $\chi = 0$; $\chi_s = 1$; hexagonal lattice ($\lambda_0 = 0.5$).

χ value will lie somewhere between 0 and 0·5, probably nearer the latter value.

Some general features of adsorption isotherms over a wide concentration range can be illustrated by using logarithmic scales for both θ and ϕ_* which highlight the behaviour in extremely dilute solutions. Such presentation[28] is shown in Fig. 5 for various chain lengths. These results show a linear Henry region followed by a pseudoplateau region. A transition concentration ϕ_*^c can be defined by extrapolation of the two linear isotherm parts. ϕ_*^c decreases exponentially with increasing chain length and when $r = 50$ the value of ϕ_*^c is so small (10^{-12}) that it does not appear within the scale shown in Fig. 5. With $r = 1000$, ϕ_*^c now reaches the ridiculously low value of 10^{-235}. The region below ϕ_*^c is the Henry region. This is the domain where the adsorbed polymer molecules behave essentially as isolated chains. Since the ϕ_*^c values are extremely low even with chains with moderate molecular weight, it becomes clear why the earlier theories that treat polymer molecules as isolated chains are not relevant in most experimental situations.

The presentation in Fig. 5 is also pertinent for the question of reversibility versus irreversibility for polymer adsorption. For high

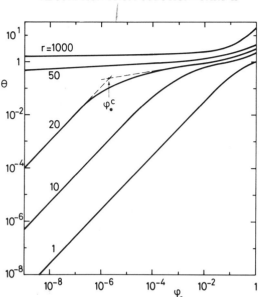

Fig. 5. Log–log presentation of adsorption isotherms of various r values. $\chi_s = 1$; $\chi = 0 \cdot 5$ hexagonal lattice ($\lambda_0 = 0 \cdot 5$).

molecular weight polymers ($r > 50$), the pseudoplateau region extends down to extremely low concentration ($\phi_*^c \sim 10^{-12}$) and this explains why one cannot easily detect any desorption upon dilution experimentally. Clearly if such extremely low concentrations can be reached desorption of the polymer may take place. However, this is not easily achieved in practice. Thus, according to Scheutjens and Fleer's theory,[28] the lack of desorption is due to the fact that the equilibrium between adsorbed and free polymer is shifted far in favour of the surface because of the high number of possible attachments per chain. Another point that emerges from Scheutjens and Fleer's theory[25–28] is the difference in shape between experimental and theoretical adsorption isotherms in the low concentration region. The experimental isotherms are usually rounded, whereas those predicted from theory are flat. This was accounted for in terms of the molecular weight polydispersity which is encountered in most practical systems. The effect of polydispersity has been explained by Cohen-Stuart et al.[30] With polydisperse polymer fractions the larger molecules adsorb preferentially over the smaller ones. At low polymer concentrations, nearly all molecular weights are adsorbed leaving only a small fraction of polymer with the lowest molecular weights in solution. As the polymer concentration is increased, the higher molecular weight fractions displace the lower ones on the surface, which are now released in solution thus shifting the molecular weight distribution of the dissolved polymer to higher values. This process continues with further increase in polymer concentration leading to a fractionation whereby the higher molecular fractions are adsorbed at the expense of the lower fractions which are released to the bulk. This process of preferential adsorption occurs at low and intermediate concentrations. However, in very concentrated solutions, monomers adsorb preferentially with respect to polymers and short chains with respect to larger ones. This is due to the fact that, in this region, the conformational entropy term dominates the free energy, disfavouring the adsorption of long chains. For a detailed analysis of this process the reader should refer to Refs 27, 30 and 31.

The dependence of adsorbed amount at various concentrations on chain length is illustrated in Fig. 6. At $\phi_* < 10^{-2}$, θ increases linearly with $\log r$ for not too short chains adsorbing from a θ-solvent. For

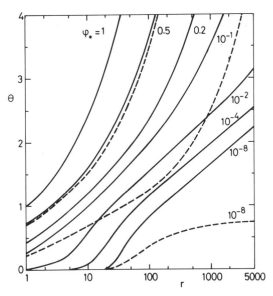

Fig. 6. Chain length dependence of the adsorbed amount at various solution concentrations. ——, $\chi = 0.5$; – – –, $\chi = 0$; $\chi_s = 1$; hexagonal lattice ($\lambda_0 = 0.5$).

more concentrated solutions the adsorbed amount increases more strongly with increase of r; indeed in this region $\theta \propto r^{\frac{1}{2}}$. For athermal solvents ($\chi = 0$), θ has a much lower dependency on r when $\phi_* < 10^{-2}$. However, in more concentrated solutions the trend of change of θ with r follows roughly that for θ-solvents.

4.1.2. Bound Fraction and Direct Surface Coverage

Figure 7 shows the variation of bound fraction p (fraction of segments in trains) and the fractional first layer coverage θ_1 (note that $\theta_1 = \phi_1$ and $p = \theta_1/\theta$) with $\log r$ at various ϕ_* values. In the Henry region ($\phi_* < \phi_*^c$), p is rather high and independent of chain length r when $r \geq 20$. In this region the molecules lie nearly flat on the surface with $p \sim 0.87\%$. At $\phi_* = 1$ (pure polymer) $p \propto r^{-\frac{1}{2}}$. This stems from the fact that $p = \theta_1/\theta$ ($\theta_1 = 1$) with $\theta \propto r^{\frac{1}{2}}$ as discussed above. The other curves for p versus $\log r$ for intermediate concentrations are between these two extreme cases. The data of Fig. 7(b) are obtained from those of

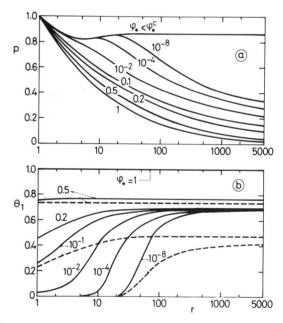

Fig. 7. (a) Bound fraction and (b) direct surface coverage as a function of chain length at various solution concentrations. ———, $\chi = 0.5$; - - -, $\chi = 0$; $\chi_s = 1$; hexagonal lattice ($\lambda_0 = 0.5$).

Fig. 6, and Fig. 7(a) by putting $\theta_1 = p\theta$. The most important features are the independence of θ_1 on r and ϕ_* for long chains, when $\phi_* \leqslant 0\cdot2$ and $\chi = 0\cdot5$. For shorter chains θ_1 increases sharply with r in a chain length region that depends on ϕ_*. In a good solvent $(\chi > 0\cdot5)$ θ_1 values are lower because of the stronger repulsion between adjacent segments. As expected, this effect is most pronounced at low solution concentrations of polymer (e.g. $\phi_* \leqslant 10^{-1}$).

4.1.3. Structure of the Adsorbed Layer

The structure of the adsorbed layer is described in terms of the segment density distribution from the surface to the bulk solution, the fraction of segments in trains, tails and loops, and the distribution of their sizes. Scheutjens and Fleer's theory[25,26] has shown how such information can be obtained for polymers near an interface. The overall segment density distribution is not easy to obtain for polymers since the r chain segments are not independent of each other. The segment density depends on the number of chains in each conformation, and is the result of contributions of all the r chain segments. Hence, the volume fraction $\phi_{i(s)}$ of segment s $(s = 1, 2, \ldots, r)$ is found from a summation of the probabilities $P(s, i, r)$ that the sth segment of an r-mer is found in the ith layer. Using this summation, Scheutjens and Fleer[25] obtained the segment density profile ϕ_i and free segment probability p_i as a function of the distance from the surface. A typical example is given in Fig. 8. For comparison the concentration profile calculated from the theory of Roe[24] is also shown in the same figure. The results of the calculations show that the segment density ϕ_i decreases continuously with increasing distance from the surface. Close to the surface ϕ_i is only slightly lower than the values predicted from the theory of Roe,[24] but at large distances the theory of Roe gives a serious underestimation for the segment density. The segment density at large distances is due to the presence of long dangling tails which have been neglected in previous theories[23] which give an exponential decay of ϕ_i with distance for homopolymers.

Scheutjens and Fleer[26] have also calculated the segment concentration profile due to tails and loops, and some typical results are shown for $r = 1000$, $\phi_* = 10^{-6}$ and $\chi = 0\cdot5$ in Fig. 9. In this example, 38% of the segments are in trains, 55·5% in loops and 6·5% in tails. This shows the difference between Scheutjens and Fleer's and earlier theories. As mentioned previously, the latter theories predict an exponential concentration profile whereas that of Scheutjens and Fleer

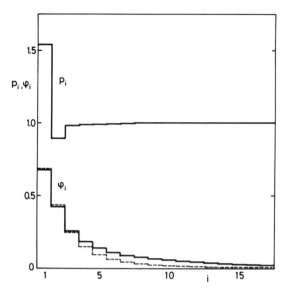

Fig. 8. Segment density profile ϕ_i and the free segment probability p_i as a function of the distance from the surface (in layer number i). Hexagonal lattice ($\lambda_0 = 0.5$); $\chi_s = 1$; $\chi = 0.5$; $r = 1000$; $\phi_* = 0.01$. $---$, concentration profile calculated from the theory of Roe.[24]

predicts that the exponential decay is only correct for the loop segments. The theory also shows the importance of tails whose segment distribution dominates the total distribution in the outer region of the adsorption layer even if tails represent only a small fraction of the total polymer segments. Only at very small distances ($i < 5$) do the loops dominate over the tails.

The influence of molar mass on the adsorbed layer thickness δ for various values of ϕ_* is shown in Fig. 10. In very dilute polymer solutions ($\phi_* \leqslant 10^{-6}$ isolated chains) δ is small and nearly independent of r. For finite concentration ($\phi_* = 10^{-2}$) and $\chi = 0.5$, δ increases linearly with $r^{\frac{1}{2}}$. This is in agreement with Hoeve's theory, in spite of his neglect of the influence of tails. However, calculation of the root mean square (rms) thickness due to loop and tail segments separately showed that both δ_1 and δ_t are proportional to $r^{\frac{1}{2}}$. For $\chi = 0$, the variation of δ with $r^{\frac{1}{2}}$ is not completely linear and lies somewhat below that for $\chi = 0.5$. In bulk polymer ($\phi_* \to 1$) δ is obviously independent of χ and χ_s.

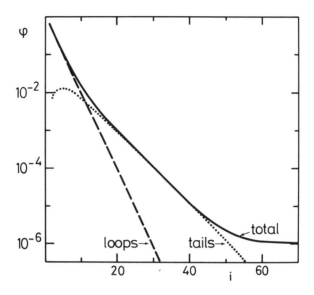

Fig. 9. Loop, tail and total segment concentration profiles according to Scheut-jens and Fleer's theory.[26] $\lambda_0 = 0 \cdot 5$; $\chi = 0 \cdot 5$; $\chi_s = 1$; $r = 1000$; $\phi_* = 10^{-6}$.

Comparison of δ with the dimension of the polymer chain in solution, namely its radius of gyration R_G, has shown that in bulk polymer δ is slightly larger than R_G while in dilute solution it is 2–3 times lower. Comparison of results of Fig. 10 with those of Fig. 6 also shows that with increasing ϕ_*, δ increases much more strongly with r than the adsorbed amount. This is due to the increasing tail fraction at high ϕ_*.

The number of trains, loops and tails, their length and the number of segments in each have also been calculated by Scheutjens and Fleer[26,28] at various ϕ_* values and an example is given for $r = 1000$ and $\chi = 0 \cdot 5$ in Table 1. The results show that when $\phi_* < \phi_*^c$ (Henry region) the chain lies virtually flat with 87% of the segments in trains, very short loops and negligible tails. The rms layer thickness is only 1·2 lattice units. In dilute solution ($\phi_* = 10^{-6}$) the picture changes significantly; the trains are shorter (and contain 38% of the segments), the loops are much larger (comprising 56% of the segments) and relatively long tails develop giving rise to a much thicker layer (4·0 lattice units).

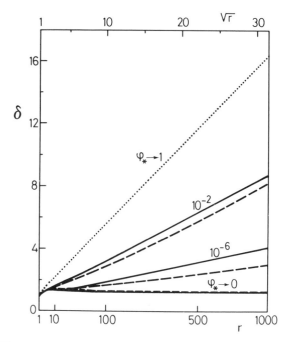

Fig. 10. The root mean square layer thickness δ (expressed as a number of lattice layers t) as a function of chain length r, at different ϕ_* values. $\chi_s = 1$. ———, $\chi = 0.5$; ———, $\chi = 0$; ·····, bulk polymer.

In semi-dilute solutions ($\phi_* = 10^{-2}$), the tails grow considerably at the expense of trains and to a lesser extent loops, giving much larger thickness (8·7 lattice units). In the limit $\phi_* \to 1$ the adsorbed polymer consists essentially of two long tails (1/3 chain length each) with a middle section of short trains and relatively long loops.

Figure 11 shows in detail how the fraction of segments in trains, in tails and in loops changes with ϕ_*. One interesting feature in Fig. 11 is the weak dependence of fraction of segments in tails on molecular weight in concentrated solutions. In dilute and semi-dilute solutions, the tail fraction is, for long chains, even independent of r. Since the number of tails per chain does not change appreciably with r, it is clear that the tail length is approximately proportional to molecular weight. Hence it is expected that the main contribution to the interaction of polymer-covered particles stems from the tails. The effect of solvent quality is rather subtle.

TABLE 1

Conformation of Adsorbed Chains with $r = 1000$, $\chi_s = 1$, $\chi = 0.5$, $\lambda_0 = 0.5$

ϕ_*	Trains			Loops			Tails			rms thickness
	Number	Length	Segments	Number	Length	Segments	Number	Length	Segments	
$<\phi_*^c$	85.7	10.1	868	84.7	1.6	131	0.4	2	1	1.2
10^{-6}	86.9	4.4	378	85.1	6.5	556	1.5	44	66	4.0
10^{-2}	64.6	4.3	278	63.6	8.0	508	1.6	131	214	8.7
1	15.7	7.5	54	14.7	20.6	303	1.9	343	643	16.4

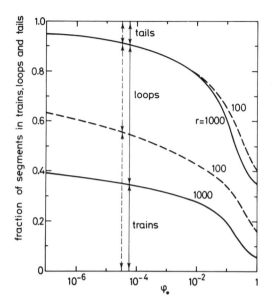

Fig. 11. Fraction of segments in trains, loops and tails as a function of ϕ_* for $r = 100$ and $r = 1000$. $\chi_s = 1$; $\chi = 0 \cdot 5$; hexagonal lattice ($\lambda_0 = 0 \cdot 5$).

4.2. Comparison of Theoretical Prediction with Experimental Results

Full quantitative comparison is not possible at present due to the lack of reliable experimental data. However, there are sufficient experimental results to show that the basic predictions of Scheutjens and Fleer's theory are essentially correct. Following the same subdivisions of the theory section, the comparison with experimental results is given below.

4.2.1. Adsorption Isotherms

The dependence of the amount of polymer adsorbed with polymer concentration and molecular weight is in broad agreement with those predicted from theory. Ideally the best test of the theoretical predictions would be provided by the use of homopolymers of very narrow molecular weight distribution adsorbed into substrates of known surface geometry. Unfortunately results obtained using such model systems are scarce. However, fairly recently van der Linden and van Leemput[32] have investigated the adsorption of narrow molecular weight fractions of poly(styrene) ($n = 500$ to 2×10^6, where n is the number of segments) onto non-porous Aerosil silica from both cyclohexane at 35°C (i.e. a θ solvent for poly(styrene)) and carbon tetrachloride (a 'good' solvent). Similarly Kawaguchi et al.[33] obtained results for poly(styrene) adsorption on silica from cyclohexane. These results showed the expected trend of sharp rise in the amount adsorbed with increasing polymer concentration, reaching a wide plateau region. Moreover, the amount adsorbed at the plateau Γ showed the expected trend of increase of Γ with increase of molecular weight. Indeed Scheutjens and Fleer[27,28] used the above results to test their theory. The expected logarithmic dependence of θ on r in θ-solvents and the weaker dependence in better solvents were both strongly corroborated. Excellent agreement between theory and experiment was demonstrated (using a value of 0·6 for χ_s) except for the oligomers which showed slight deviation from the theoretical curves.

Results for polymer adsorption from aqueous solution using homopolymers with narrow molecular weight distributions are very scarce. Most studies with water-soluble polymers were carried out using commercial materials which are poorly characterised. Perhaps the most thorough investigations were those obtained using poly(vinyl alcohol) (PVA).[34–36] Since this polymer is invariably polydisperse and contains some residual unhydrolysed vinyl acetate residues, it is clearly unsuitable for quantitative tests of theoretical predictions. Garvey et al.[34] attempted to overcome the former problem by using gel permeation chromatography to prepare narrow molecular weight fractions of the polymer. The fractions were then used for a study of adsorption on monodisperse polystyrene latex particles.[34,35]

The results showed the expected trend of increase of adsorption with increase of molecular weight, and the adsorbed value at the plateau showed a linear increase with square root of molecular weight. Moreover in later experiments[36,37] it was shown that adsorption increased with increase of temperature or addition of high concentration of electrolyte, both of which result in a reduction of solvency for the PVA chain. All these trends are in qualitative agreement with the theory demonstrating its utility in describing the properties of practical systems.

The effects of polydispersity of the polymer on the shape of the adsorption isotherm have been demonstrated by Cohen-Stuart et al.[30] for the adsorption of poly(vinyl pyrrolidone) (PVP) on silica from aqueous solution. Two narrow molecular fractions of PVP with $M =$

1440 and $1 \cdot 0 \times 10^6$ were used (heterodispersity index $M_v/M_n < 1 \cdot 5$). These samples were then mixed to give a heterodisperse sample. The narrow molecular weight fractions gave the expected 'flat' type isotherm with a sharp rise at low concentrations, whereas the mixtures gave either a 'rounded' isotherm or a discontinuous pattern that is predicted from the theory.[30] Although quantitative agreement between theory and experiment was not obtained (possibly due to the fact that the fractions used were not ideally monodisperse), the trends observed were again in good qualitative agreement with theoretical predictions.

4.2.2. Bound Fraction

The relationship between bound fraction p and r has been confirmed by Scheutjens and Fleer[28] for the adsorption of poly(styrene) on silica.[32,33] The latter results showed a reduction of p with increase of M reaching a limiting value of $0 \cdot 22$ at high M. However, values of p obtained for adsorption of poly(styrene) on silica reported in the literature were not always the same. For example, Joppien[38] found a somewhat higher value ($\sim 0 \cdot 34$) for poly(styrene) with $M \sim 2 \cdot 5 \times 10^{-5}$ adsorbed onto non-polar silica from carbon tetrachloride, whereas Robb and Sharples[39] reported lower values for p ($< 0 \cdot 1$). The large discrepancy between different authors may be due to the difference in the number of surface hydroxyl groups on the various silicas used.

Several results are available in the literature for the bound fraction of other polymers, although silica has mostly been chosen as a convenient substrate. For example, Robb and Smith[40,41] measured the bound fraction of poly(vinyl pyrrolidone) (PVP) and poly(methyl methacrylate) (PMMA) on silica in chloroform using electron paramagnetic resonance. The results showed that PVP adopts a flat conformation at low coverage (high p) changing to loop–train conformation (large loops and low p) at high coverage. In contrast PMMA results suggest that a high p is retained up to saturation coverage, thus indicating a flat conformation over the range of surface coverage studied. Clark et al.[42] showed that p for PVP on silica increases with reduction of solvency, in line with theoretical predictions.

The bound fraction of poly(ethylene oxide) (PEO) adsorbed onto silica from carbon tetrachloride was shown by Killmann[43] (using microcalorimetry) to decrease from $0 \cdot 5$ to $0 \cdot 2$ with increasing coverage. On metal surfaces, Killmann et al.[44] have also shown (using ellipsometric, IR and calorimetric measurements) that PEO adopts a flatter configuration than either poly(styrene) or PMMA in a number of solvents.

These results on the bound fraction clearly demonstrate that flexible macromolecules like PEO and PMMA can interact strongly with the substrate (through hydrogen-bonding) leading to relatively high p values.

4.2.3. Adsorbed Layer Thickness and Segment Density Distribution
The square root dependence of layer thickness on molecular weight has been demonstrated by several authors. For example Killmann et al.[44] showed such a dependence for the adsorption of polystyrene on metal surfaces. Similar results were obtained by Takahashi et al.[45] for poly(styrene) adsorbed on chrome surface from cyclohexane at 35°C. The latter authors also showed that the polymer layer thickness rises quite steeply with increasing bulk polymer concentration and then levels off to a constant value at high polymer concentration. Again the results are in semi-quantitative agreement with the theory of Scheutjens and Fleer.[26,27]

Several results were reported in the literature for the adsorbed layer thickness of water-soluble polymers, which in the majority of cases have been obtained from hydrodynamic measurements (see below). The most systematic investigations were those performed by Tadros and co-workers[35-37] for PVA on poly(styrene) latex particles. Initially the adsorbed layer thickness was measured on small (60 nm) poly(styrene) latex particles using an ultracentrifugation technique.[35] These results showed a square root dependence of δ on molecular weight. Later, however, the layer thickness was measured as a function of particle radius a and was shown to increase with increase of a (although the adsorbed amount Γ did not change with particle radius) and this was ascribed to a geometric packing effect of the adsorbed macromolecules. When the equivalent flat surface value for δ, i.e. δ_{eff}, was calculated, it was found to be proportional to $M^{1.1}$ rather than $M^{0.5}$ found for the small particle case. Later, the effect of solvency on δ was also investigated[36] and it was shown that δ decreases significantly with reduction of solvency as expected.

Experimental results on the segment density distribution from the surface are very scarce. Recently Barnett and co-workers[46-48] obtained some data for PVA on deuterated poly(styrene) latex using neutron scattering. The results obtained for PVA with $M_w = 37\ 000$ are shown in Fig. 12. This shows a monotonic decay of the segment density distribution $\rho(z)$ with distance z from the surface and several regions may be distinguished. Close to the surface $(0 < 2 > 3\ nm)$, the decay in

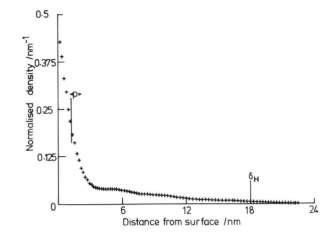

Fig. 12. Plot of $\rho(z)$ against z for 37 000 molecular weight PVA adsorbed on deuterated PS latex in D_2O/H_2O.

$\rho(z)$ is rapid and, assuming the thickness of 1·3 nm for the bound layer, p was calculated to be 0·1. In the middle region, the segment density shows a shallow maximum and a slow decay which extends up to the hydrodynamic radius, δ_H, at 18 nm. The latter is determined by the longest tails and is ~2·5 times the radius of gyration in bulk solution (7·2 nm). Thus the slow decay of $\rho(z)$ with z at long distances is in qualitative agreement with Scheutjens and Fleer's theory[26,27] which predicts the presence of long tails. Moreover, the existence of a shallow maximum at intermediate distances suggests that the observed segment density is a summation of a fast monotonic decay due to loops and trains together with the segment density for tails which shows a maximum density away from the surface. The latter maximum was clearly observed for a sample which had PEO terminally grafted to a deuterated poly(styrene) latex (i.e. where the configuration is represented by tails only).

5. EXPERIMENTAL METHODS FOR MEASUREMENT OF POLYMER ADSORPTION AND CONFIGURATION

For the full description of the adsorption of polymers from solution, it is necessary to measure four important quantities: the amount of adsorption per unit area (Γ), the fraction of bound segments (p), the adsorbed layer thickness δ, and segment density distribution $\rho(z)$. It is essential to know how these parameters change with polymer concentration, degree of polymerisation (i.e. molecular weight) and solvency (as determined by the interaction parameter χ). A brief summary of how each of these quantities can be measured is given below.

5.1. Amount of Polymer Adsorbed (Γ)

The determination of adsorption isotherms, i.e. Γ as a function of C_2 (the 'equilibrium' concentration in solution) is fairly well established for physically adsorbed polymers. Basically, one determines the change in polymer concentration, ΔC_2, in the bulk solution phase before and after equilibration with the solid particles (of known surface area) in question. It is essential to develop analytical techniques that are capable of measuring low concentrations (~ppm) in order to establish the initial steeply rising part of the isotherm. Sensitive analytical techniques such as radiotracer, IR absorption and differential interferometry may be applied, and in some cases calorimetric methods based on complex formation between the polymer and specific reagents can also be adapted for determination of low polymer concentration.

Although simple in principle, in practice some care is required to establish reliable isotherms. This has to do with the problem of 'irreversibility' and the adsorption/desorption kinetics. As mentioned before the apparent 'irreversibility' of polymer adsorption results from the 'multipoint' attachment to the surface and this is a time-dependent phenomenon. This immediately raises the question of the validity of application of equilibrium thermodynamics, e.g. the Gibbs adsorption isotherm. The attainment of equilibration in studying polymer adsorption represents another problem. As a result of the low diffusion coefficient of polymer molecules in solution, and the finite time taken for a polymer molecule to adopt its steady-state adsorbed conformation, long equilibration times (hours to days) are needed. This is particularly the case at high coverage and when the polymer is polydisperse. The smaller molecules which have higher diffusion coefficients will adsorb first and these are gradually replaced by the preferentially adsorbed larger molecules. Several studies have given experimental evidence for such a preference.[49-51]

5.2. Polymer Bound Fraction (p)

Several techniques are available for measurement of the fraction of segments of an adsorbed polymer that are in direct contact with the surface. The most direct methods for assessing p are spectroscopic methods namely infra-red (IR) electron spin resonance (EPR, also referred to as ESR) and nuclear magnetic resonance (NMR). The IR method depends on measuring the shift in some absorption peak for a polymer adsorbed from solution.[44,52] ESR and NMR methods depend on the assumption that for an adsorbed homopolymer the segments in trains have a lower mobility, i.e. a longer rotational correlation time than those in loops. In the ESR method[41] it is necessary to use a chemical spin label, e.g. a nitroxide group. The assumption is made that the introduction of a small number of such groups into the polymer does not affect the adsorption, particularly when such groups are randomly distributed. In this respect the NMR method has a major advantage since it does not require attachment of a label and hence requires no such assumption. A pulsed NMR technique has been recently applied by Barnett et al.[46] for the estimation of p.

An indirect method for the estimation of p is microcalorimetry. Basically, one compares the measured enthalpy of adsorption per molecule with the enthalpy of adsorption per segment.[43] This ratio gives the number of segments per molecule in contact with the surface. The enthalpy of adsorption per segment is assumed to be equivalent to that for a small molecule of equivalent structure. The latter can be determined, for example, from the measured heat of wetting of such a small molecule on the same surface. Although one may question the validity of this basic assumption the major problem of this technique is disentangling the enthalpy of adsorption from the other factors which contribute to the measured enthalpy value. The latter is the net value from a number of contributions, e.g. solvent desorption, segment adsorption and overall reconformation.

Recently Cohen-Stuart et al.[53] compared the values of p obtained using IR, NMR and microcalorimetry for PVP adsorption on silica. The p values obtained from IR agree rather poorly with theoretical data, whereas the EPR and NMR agree well with each other and with theoretical predictions. The microcalorimetry also yields reasonable results, but only for relatively short polymer chains. Thus it seems that EPR and NMR give the most reliable methods for estimation of p.

5.3. Adsorbed Layer Thickness and Segment Density Distribution

Three direct methods can be applied for determination of adsorbed layer thickness of adsorbed polymers: ellipsometry, attenuated total reflection (ATR) and neutron scattering. Both ellipsometry and ATR[54,55] depend on the differences in refractive indices between the substrate, the adsorbed layer and the bulk solution, and require an optically flat reflecting surface. Ellipsometry[55] is based on the principle that light undergoes a change in polarisability when it is reflected at a flat surface (whether covered or uncovered with a polymer layer). As a result of polymer adsorption both the amplitude ratio of the parallel to perpendicular components of the reflected beam and the phase shift, Δ, change. This relative change is used to calculate the thickness of the adsorbed layer and its refractive index. The latter can be used to calculate the average segment concentration in the adsorbed layer. In ATR one measures the absolute values of the two reflection coefficients (the parallel and perpendicular components). Both ellipsometry and ATR suffer from the disadvantage that they need optically flat reflecting surfaces which makes them less attractive for most practical systems.

In neutron scattering the adsorbed layer thickness can be measured directly from the scattering due to the adsorbed layer when the scattering length density of the particle is matched to that of the medium (the so-called 'contrast matching' method). This method has been recently used by Cebula et al.[56] for measuring the thickness of surfactants adsorbed on poly(styrene) latex. As will be seen later, the neutron-scattering method can be applied not only to obtain δ but also to determine the segment density distribution $\rho(z)$.[46–48]

For measurement of adsorbed layers in practical systems, hydrodynamic methods are commonly used. These are based on measuring the reduced viscosity, sedimentation coefficient, diffusion coefficient and electrophoretic mobility of the particles with and without adsorbed polymer layers. All these methods give a hydrodynamic thickness δ_H which, although different from the 'real' thickness (see Fig. 12), is a useful parameter in connection with stability studies.

The viscosity method[57] depends on measuring the increase in effective volume fraction of the particles, ϕ, in the dispersion as a result of the presence of the adsorbed layer. ϕ can be obtained from intrinsic viscosity measurements using the well-known Einstein equation i.e.:

$$\eta_{sp}/\phi = [\eta] + k'[\eta]^2\phi \qquad (7)$$

where η_{sp} is the specific viscosity of the dispersion ($\eta_{sp} = \eta_{rel}-1$), k' is the Huggins constant and $[\eta]$ is the intrinsic viscosity which for a dilute suspension of rigid non-interacting spheres is equal to 2·5. The presence of the adsorbed layer increases the effective volume fraction of the particle, i.e. $\phi' = f\phi$. Assuming $[\eta]$ and k' to be unchanged when the adsorbed polymer is present on the particles, one may write:

$$(\eta'_{sp}/\phi) = [\eta]f + k'[\eta]^2\phi \tag{8}$$

where η'_{sp} is the specific viscosity in the presence of adsorbed layer. Thus, from a plot of (η'_{sp}/ϕ) versus ϕ and extrapolation to $\phi = 0$, one obtains $[\eta]f$ from the intercept. For spheres with average radius $\bar{a}$, δ_H is related to f by the expression:

$$\delta_H = \bar{a}(f^{\frac{1}{3}} - 1) \tag{9}$$

Since $[\eta]f$ is obtained from extrapolation to $\phi = 0$, the effect of interparticle interaction is minimised. However, it is necessary to reduce the primary electroviscous effect by addition of electrolyte.

The sedimentation method depends on measuring the sedimentation coefficient (e.g. using an ultracentrifuge) of the particles (extrapolated to zero concentration) in the presence of adsorbed polymer.[34,35] Assuming the particles obey Stokes' law, the sedimentation coefficient S'_0 in the presence of an adsorbed polymer layer is given by:

$$S'_0 = \frac{(4/3)\pi a^3(\rho - \rho_s) + (4/3)\pi[(a + \delta_H)^3 - a^3](\rho_s^{ads} - \rho_s)}{6\pi\eta(a + \delta_H)} \tag{10}$$

where η is the viscosity of the medium, ρ and ρ_s are the mass density of the solid and solution phase, respectively, and ρ_s^{ads} is the average mass density of the interfacial region contained in the adsorbed polymer. ρ_s may be obtained from the average mass concentration C_2^s for the polymer in the adsorbed layer:[35]

$$C_2^s = \frac{4\pi a^2\Gamma_2^s}{(4\pi/3)(a + \delta)^3 - (4\pi/3)a^3} \tag{11}$$

where Γ_2^s is the mass of polymer adsorbed per unit area. The density of a dilute polymer solution is related to that of the solvent by:

$$\rho_s = \rho_0 + kC_2 \tag{12}$$

where ρ_0 is the density of the solvent and k is an empirical constant. Assuming that eqn (12) applies to the interfacial region and that $\rho_s = \rho_0$

(i.e. C_2 is small), then eqns (10)–(12) may be combined to give:

$$S_0' = \frac{2a^3(\rho - \rho_s) + 6a^2\Gamma_2^s}{9\eta(a + \delta_H)} \tag{13}$$

from which δ_H may be calculated from a knowledge of Γ_2^s for the corresponding C_2 value. Thus, the ultracentrifuge technique requires knowledge of the adsorption isotherms.

A relatively simple sedimentation method for determination of δ_H, which requires fairly cheap equipment, is the slow-speed centrifugation developed by Garvey et al.[34] Basically, a stable monodisperse dispersion is slowly centrifuged at low g values ($\sim$50 g) to form a close-packed (hexagonal or cubic) lattice in the sediment. From a knowledge of the volume fraction and packing mode, the distance of separation between the centres of the two particles may be calculated. This is assumed to be equal to the diameter of the polymer-coated particles a ($= a_0 + \delta_H$). Thus for hexagonal close-packed particles (packing fraction $= 0\cdot7405$):

$$a = a_0 + \delta_H = \left(\frac{0\cdot7405\,V\rho_1 a_0^3}{W}\right) \tag{14}$$

where V is the total sediment value and W is the weight of particles. Thus δ_H can be calculated using eqn (14) from a knowledge of V.

The most convenient and rapid method for measuring the adsorbed layer thickness is from the particle diffusion coefficient D which can be obtained from quasi-elastic light scattering that is usually referred to as intensity fluctuation spectroscopy or photon correlation spectroscopy. This method depends on the measurement of the diffusion coefficient of the particles by measuring the intensity fluctuation of scattered light as they undergo Brownian motion. A full detailed analysis of the theory has been given by Pusey.[58] When a light beam passes through a colloidal dispersion, an oscillating dipole moment is induced in the particles, thus re-radiating the light. Due to the random positions of the particles, the intensity of the scattered light will, at any instant, appear as random diffraction or a 'speckle' pattern. As the particles undergo Brownian motion, the random configuration of the speckle pattern changes. The intensity at any one point in the pattern will, therefore, fluctuate such that the time taken for an intensity maximum to become a minimum (i.e. the coherence time) corresponds approximately to the time required for a particle to move one wavelength.

Using a photomultiplier of active area about the size of the diffraction maximum (i.e. approximately one coherence area), this intensity fluctuation can be measured. A digital correlator is used to measure the photocount (or intensity) correlation function of the scattered light. The photocount correlation function is given by the equation:

$$G^{(2)}(\tau) = B\{1 + \gamma^2[g^{(1)}(\tau)]^2\} \tag{15}$$

where τ is the correlation delay time. The correlator computes $G^{(2)}(\tau)$ for many values of τ. B is the background value to which $G^{(2)}(\tau)$ decays at long delay time. $[g^{(1)}(\tau)]$ is the normalised correlation function of the scattered electric field and γ is a constant ~ 1.

For monodisperse non-interacting particles:

$$[g^{(1)}(\tau)] = \exp -(\Gamma\tau) \tag{16}$$

where Γ is the decay rate or inverse coherence time, which is related to the translational diffusion coefficient D by the equation:

$$\Gamma = DK^2 \tag{17}$$

where K is the magnitude of the scattering vector, given by the equation:

$$K = \frac{4n}{\lambda_0} \sin (\theta/2) \tag{18}$$

where λ_0 is the wavelength of the incident light *in vacuo*, n is the refractive index of the solution and θ is the scattering angle.

The particle radius can be calculated from D using the Stokes–Einstein equation:

$$a = \frac{\mathbf{k}T}{6\pi\eta_0 D} \tag{19}$$

where $\mathbf{k}$ is the Boltzmann constant, T is the absolute temperature and η_0 is the viscosity of the continuous phase. For a polymer-coated particle $a = a_0 + \delta_H$. Thus, by carrying out the measurement in the presence and absence of adsorbed layer, one is able to estimate δ_H. The only limitation of the method is the relative value of δ_H/a. Since the accuracy of the method is $\sim 1\%$, it is necessary to have an adsorbed layer of $\sim 10\%$ of the value of the radius of the particles. Moreover, it is necessary to use monodisperse particles. Thus this method can be applied to the plateau region of the isotherm with relatively small and monodisperse particles.

Measurement of the electrophoretic mobility can also be used to measure the adsorbed layer thickness.[59] From the electrophoretic mobility the zeta potential (ζ), i.e. the electrostatic potential (with respect to the bulk solution phase), at the 'slipping plane' of the particle is calculated. For a particle plus an adsorbed polymer, the distance of this slipping plane from the solid surface is δ_H, the hydrodynamic thickness. However, the calculation of zeta potential from electrophoretic mobility is not necessarily straightforward, particularly under conditions where double layer relaxation effects are significant. Moreover, it is necessary to use a very dilute dispersion, e.g. in microelectrophoresis. It is also necessary to know the potential in the Stern plane, ψ_d, since:

$$\tanh\left(\frac{e\zeta}{4kT}\right) = \tanh\left(\frac{e\psi_d}{4kT}\right)\exp - \kappa(\delta_H - \Delta) \qquad (20)$$

where e is the electronic charge and Δ is the thickness of the Stern layer. Thus to estimate δ_H from eqn (20) it is necessary to know ψ_d and Δ. The latter can usually be set equal to the diameter of the (hydrated) counterion, but estimation of ψ_d is not straightforward, unless the diffuse double layer charge density σ_d is known, since:

$$\sigma_d = (8\varepsilon\varepsilon_0 C_e kT)^{\frac{1}{2}}\sinh\left(\frac{e\psi_d}{2kT}\right) \qquad (21)$$

In the absence of specific adsorption σ_d is equal to the surface charge σ_0 and if the latter can be determined independently, e.g. from potentiometric titration, ψ_d can be calculated, thus allowing δ_H to be obtained from eqn (20). The above procedure has been applied for silver iodide particles coated with PVA by Fleer et al.[59] where surface charge–surface potential curves were available. However, apart from this limitation to a system where σ_0 can be established independently, the method is subject to uncertainties in view of the various assumptions that have to be made. One of these assumptions is that the diffuse double layer is not perturbed by polymer segments in loops and tails. This is clearly not always the case. Moreover, the second assumption of equality of ψ_d and ψ_0 is not always justified unless specific adsorption is absent.

As mentioned above, thickness measurements only give qualitative information on the structure of the adsorbed polymer layer. For a quantitative description of the conformation of a macromolecule at the solid–solution interface, it is necessary to obtain the segment density distribution $\rho(z)$ and the loop–train and tail distributions. Both the radius of gyration of individual adsorbing polymer molecules and $\rho(z)$ can be obtained from small-angle neutron scattering. This depends on the distribution of scattering centres in the sample.[47] Different nuclei vary substantially in their effectiveness to scatter neutrons. In particular, isotopic variation enables molecular regions of interest to be viewed against a weakly scattering background. For example, for an adsorbed polymer on a deuterated latex particle in aqueous solution, scattering from the adsorbed layer can be detected by contrasting out the latex particle with a certain ratio of H_2O/D_2O as solvent. In this case the scattering intensity $I(O)$ at zero scattering angle θ, is given by:

$$I(O) = (\rho_m V - \rho_s V)^2 \qquad (22)$$

where ρ_m is the mean scattering length density of particle plus its adsorbed layer, ρ_s is the scattering length density of the solvent and V is the total volume of the particle plus its adsorbed layer. V is obtained from a plot of $\sqrt{[I(O)]}$ versus ρ_s and δ is then obtained from the relationship:

$$V = (4/3)\pi(a + \delta)^3 \qquad (23)$$

Moreover, by measuring the scattering intensity $I(Q)$ as a function of scattering vector Q (where $Q = (4\pi/\lambda)\sin(\theta/2)$), $\rho(z)$ can be obtained by Fourier transform of the $I(Q)$ data for dilute dispersions of monodisperse non-interacting particles. The details of the calculation have been given previously[47,48] and an example of $\rho(z)$ versus z for PVA ($M = 37\,000$) adsorbed on poly(styrene) latex is given in Fig. 12.

ACKNOWLEDGEMENTS

The author is indebted to Drs G. Fleer and B. Vincent for reading the manuscript and making some improvements to the text.

REFERENCES

1. Patat, F., Killmann, E. and Schliebener, C., *Adv. Polymer Sci.*, **3,** 332 (1964).
2. Stromberg, R. R., In *Treatise on adhesion and adhesives*, Vol. 1, R. I. Patrick (ed.), Marcel Dekker, New York (1967), p. 69.

3. Silberberg, A., *Pure Appl. Chem.*, **26**, 583 (1971).
4. Fontana, B. J., In *The chemistry of biosurfaces*, Vol. 1, M. L. Hair (ed.), Marcel Dekker, New York (1971), p. 83.
5. Ash, S. G., In *Colloid Science*, Vol. 1, The Chemical Society, London (1973), p. 103.
6. Vincent, B., *Adv. Colloid Interface Sci.*, **4**, 193 (1974).
7. Lipatov, Yu. S. and Sergeeva, L. M., *Adsorption of polymers*, John Wiley and Sons, New York (1974).
8. Tadros, Th. F., In *The effect of polymers on dispersion properties*, Th. F. Tadros (ed.), Academic Press, London (1982), pp. 1–38.
9. Vincent, B. and Whittington, S. G., In *Surface and colloid science*, Vol. 12, E. Matijevic (ed.), Plenum Press, New York (1982), pp. 1–117.
10. Takahashi, A. and Kawaguchi, M., *Adv. Polymer Sci.*, **46**, 1 (1982).
11. Jenckel, E. and Rumbach, R., *Z. Elecktrochem.*, **55**, 612 (1952).
12. Flory, P. J., *Principles of polymer chemistry*, Cornell University Press, New York (1953).
13. Simha, R., Frisch, H. L. and Eirich, F. R., *J. Phys. Chem.*, **57**, 584 (1953).
14. Frisch, H. L. and Simha, R., *J. Chem. Phys.*, **24**, 652 (1956); **27**, 702 (1957).
15. Silberberg, A., *J. Phys. Chem.*, **66**, 1872, 1884 (1962); *J. Chem. Phys.*, **46**, 1105 (1967).
16. DiMarzio, E. A., *J. Chem. Phys.*, **42**, 2102 (1965).
17. DiMarzio, E. A. and McCrackin, F. L., *J. Chem. Phys.*, **43**, 539 (1965).
18. Hoeve, C. A. J., DiMarzio, E. A. and Peyser, P., *J. Chem. Phys.*, **42**, 2558 (1965).
19. Rubin, R. J., *J. Chem. Phys.*, **43**, 2392 (1965); *J. Res. Nat. Bur. Standards Sect. B*, **70**, 237 (1966).
20. Roe, R. J., *J. Chem. Phys.*, **43**, 1591 (1965); **44**, 4264 (1966).
21. Motomura, K. and Matuura, R., *J. Chem. Phys.*, **50**, 1281 (1969); Motomura, K., Sekita, A. and Matuura, R., *Bull. Chem. Ac. Jap.*, **44**, 1243 (1971); Motomura, K., Moroi, Y. and Matuura, R., *Bull. Chem. Soc. Jap.*, **44**, 1248 (1971).
22. Silberberg, A., *J. Chem. Phys.*, **48**, 2835 (1968).
23. Hoeve, C. J. A., *J. Polymer Sci. C*, **30**, 361 (1970); **34**, 1 (1971).
24. Roe, R. J., *J. Chem. Phys.*, **60**, 4192 (1974).
25. Scheutjens, J. M. H. M. and Fleer, G. J., *J. Phys. Chem.*, **83**, 1619 (1979).
26. Scheutjens, J. M. H. M. and Fleer, G. J., *J. Phys. Chem.*, **84**, 178 (1980).
27. Scheutjens, J. M. H. M. and Fleer, G. J., In *The effect of polymers on dispersion properties*, Th. F. Tadros (ed.), Academic Press, London (1982), p. 145.
28. Scheutjens, J. M. H. M. and Fleer, G. J., *Adv. Colloid Interface Sci.*, **16**, 341 (1982).
29. DiMarzio, E. A. and Rubin, R. J., *J. Chem. Phys.*, **55**, 4318 (1971).
30. Cohen-Stuart, M. A., Scheutjens, J. M. H. M. and Fleer, G. J., *J. Polymer Sci. Polymer Phys. Ed.*, **18**, 559 (1980).
31. Hesselink, F. Th., *J. Phys. Chem.*, **75**, 65 (1971).
32. van der Linden, C. and van Leemput, R., *J. Colloid Interface Sci.*, **67**, 48 (1978).

33. Kawaguchi, M., Hayakawa, K. and Takahashi, A., *Polymer J.*, **12**, 265 (1980).
34. Garvey, M. J., Tadros, Th. F. and Vincent, B., *J. Colloid Interface Sci.*, **49**, 57 (1974).
35. Garvey, M. J., Tadros, Th. F. and Vincent, B., *J. Colloid Interface Sci.*, **55**, 440 (1976).
36. Van der Boomgaard, Th., King, T. A., Tadros, Th. F., Tang, H. and Vincent, B., *J. Colloid Interface Sci.*, **61**, 68 (1978).
37. Tadros, Th. F. and Vincent, B., *J. Colloid Interface Sci.*, **72**, 505 (1979).
38. Joppien, G. R., *Makromol. Chem.*, **175**, 1931 (1974); **176**, 1129 (1975).
39. Robb, I. D. and Sharples, N., Private communication in Ref. (9).
40. Robb, I. D. and Smith, R., *Polymer*, **18**, 500 (1977).
41. Robb, I. D. and Smith, R., *Eur. Polymer. J.*, **10**, 1005 (1974).
42. Clarke, A. J., Robb, I. D. and Smith, R., *J. Chem. Soc. Faraday Trans. I*, **72**, 1489 (1976).
43. Killmann, E., *Polymer*, **17**, 864 (1976).
44. Killmann, E., Eisenlauer, J. and Korn, M. J., *Polymer Sci. Polymer Symp.*, **61**, 413 (1977).
45. Takahashi, A., Kawaguchi, M., Hirota, H. and Kato, T., *Macromolecules*, **13**, 884 (1980).
46. Barnett, K. G., Cosgrove, T., Vincent, B., Burgess, A. N., Crowley, T. L., King, J., Turner, J. D. and Tadros, Th. F., *Polymer*, **22**, 283 (1981).
47. Barnett, K. G., Cosgrove, T., Crowley, T. L., Tadros, Th. F. and Vincent, B., In *The effect of polymers on dispersion properties*, Th. F. Tadros (ed.), Academic Press, London (1982), p. 183.
48. Cosgrove, T., Crowley, T. L., Vincent, B., Barnett, K. G. and Tadros, Th. F., *Faraday Symp.*, 101 (1981).
49. Felter, R. E., Mayer, E. S. and Ray, L. M., *J. Polymer Sci.*, **B7**, 529 (1969).
50. Felter, R. E. and Ray, L. M., *J. Colloid Interface Sci.*, **32**, 349 (1970).
51. Howard, G. J. and Woods, S. J., *J. Polymer Sci. A-2*, **10**, 1023 (1972).
52. Fontana, B. J. and Thomas, J. R., *J. Phys. Chem.*, **65**, 480 (1961).
53. Cohen-Stuart, M. A., Fleer, G. J. and Bijsterbosch, B. H., *J. Colloid Interface Sci.*, **90**, 310, 321 (1982).
54. Peyser, P. and Stromberg, R. R., *J. Phys. Chem.*, **71**, 2066 (1967).
55. Stromberg, R. R., Passaglia, E. and Tutas, D. J., *J. Res. Nat. Bur. Standards*, **67**(A), 431 (1965).
56. Cebula, D., Thomas, R. K., Harris, N. H., Tabany, J. and White, J. W., *Faraday Disc. Chem. Soc.*, **65**, 76 (1978).
57. Doroszkowski, A. and Lambourne, R., *J. Colloid Interface Sci.*, **26**, 214 (1968).
58. Pusey, P. N., In *Industrial polymers: characterisation by molecular weight*, J. H. S. Green and R. Dietz (ed.), Transcripta Books, London (1979).
59. Fleer, G. J., Koopal, L. K. and Lyklema, J., *Kolloid Z. u. Z. Polymere*, **250**, 689 (1972).

Chapter 5

The Stability of Polymer Latices

R. BUSCALL

Corporate Bioscience and Colloid Laboratory, ICI PLC, Runcorn, UK

and

R. H. OTTEWILL

School of Chemistry, University of Bristol, UK

1. THE NATURE OF POLYMER LATEX PARTICLES

The majority, if not all, polymer latex particles are composed of a large number of polymer chains, with the individual chains having molecular weights in the range 10^5-10^7. Depending on the arrangement of the chains within the particle it can be amorphous, crystalline, rubbery or glassy. Many homopolymer latices, e.g. poly(styrene) and poly(vinyl acetate), have a well-defined glass transition temperature. Typically, those latices which remain swollen with monomer, in which the polymer is soluble, until the last stages of conversion assume a spherical shape. Poly(tetrafluoroethylene) (PTFE) latices on the other hand, in which the monomer is not soluble, form crystalline particles and consequently have a non-spherical shape. The physical state of the particles can be important in close-range interactions and in drying. For example, with soft particles, coalescence of the particles can occur more easily, leading to film formation in the dry state, whereas with hard particles their individuality is retained in this state.

The colloidal behaviour of a dispersion of latex particles is directly related to the surface properties of the particles and these in turn depend on the preparative methods employed for the synthesis of the particles. In aqueous-based polymerisations, both in the presence and

141

absence of emulsifier, surface groups are formed on the particle surface which derive from the initiator used. Some typical examples are:

$$\text{Weak acid} \qquad -C\!\!\underset{\displaystyle OH}{\overset{\displaystyle O}{<}}$$

from hydrogen peroxide, bisazocyanopentanoic acid or side reactions following the use of persulphate.

$$\text{Strong acid} \qquad -O-SO_3^-$$

from persulphate.

$$\text{Non-ionic} \qquad -OH$$

from hydrogen peroxide or hydrolysis of sulphate groups.

The use of mixed initiators can lead to the formation of amphoteric surfaces.

In addition when emulsifiers are used in the preparation these can remain adsorbed onto the particle surface and in some cases direct grafting to the surface occurs. Non-ionic polymers and polyelectrolytes can also be either adsorbed on to or grafted on to the particle surface.

When the particles are dispersed in a medium of high relative permittivity, for example water, the acidic and basic groupings exist on the surface in an ionised form, the extent of ionisation depending on the pK_a and pK_b values of the surface groups and the pH of the dispersion medium.

In many cases the end-groups or other surface moieties generated in the polymerisation process will not occupy the total surface area of the particle. Some parts of the surface will be occupied by polymer molecules, e.g. poly(styrene) and poly(tetrafluoroethylene), and hence will be essentially hydrophobic. The influence that these patches have on the latex properties will in turn depend on the range of influence of the charged sites (see Section 2.13) and hence it is possible to have latex particles which are hydrophilic at low salt concentrations but hydrophobic at high salt concentrations.

Clearly, it is possible to produce in practice a wide variety of latex particles and a schematic illustration of some types is given in Fig. 1. For systems where molecular extension occurs into the dispersion medium, as illustrated in Figs 1(b), 1(c) and 1(d), the steric configurations in the now three-dimensional outer layer have to be taken into account in a consideration of the particle properties.

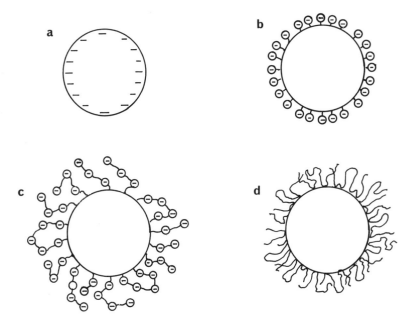

Fig. 1. Schematic illustration of some possible types of latex particle: (a) smooth, (b) short ionic chains, (c) long ionic chains, (d) long non-ionic chains.

From this qualitative description of latex particles we can recognise the origin of the basic interparticle forces that have to be taken into account in understanding the stability behaviour of dispersions in both aqueous and non-aqueous dispersion media. These can be summarised as:

(1) *Electrostatic effects:* arising from ionised surface groups—usually repulsive but particles having opposite charges will undergo electrostatic attraction; they depend on electrolyte concentration.

(2) *Steric effects:* arising from the geometry and conformation of adsorbed or grafted molecules on the surface—can also include charge effects when the molecules are polyelectrolytes. These can be repulsive or attractive and are often temperature-dependent and can depend on electrolyte concentration.

(3) *Attractive effects:* arising from long-range dispersion forces when, essentially, the density and polarisability of the molecules

constituting the particles are different from those of the medium. Generally electrolyte- and temperature-independent.

(4) *Solvation effects:* arising from the organisation of solvent molecules near an interface or a macromolecule in a different manner from that of the bulk phase.

2. ELECTROSTATICALLY STABILISED LATICES

2.1. The Electrostatics of a Latex Surface–Medium Interface

In Fig. 1(a) a smooth latex surface is illustrated with a number of negative charges located at the interface. Experimentally, these charges are directly accessible by conductimetric titration, usually after treatment of the latex with a mixed-bed ion-exchange resin to convert it into the H^+ or OH^- form. The quantity obtained is the surface charge density, σ_0, defined by:

$$\sigma_0 = N_c ev \qquad (1)$$

where N_c is the number of charged sites per cm^2 and v is their valency. e is the fundamental charge on the electron. The values obtained, in general, lie in the range $0 \cdot 5$–$10 \ \mu C \, cm^{-2}$.

As a consequence of the surface charge σ_0, the surface acquires a surface potential, ψ_0, relative to the bulk solution (earth). This can be positive or negative depending on the nature of the surface groups. In order to maintain the condition of electroneutrality, the charge on the surface has to be compensated by an equal and opposite charge in the solution phase. Thus ions of opposite charge to the interfacial charge (counterions) are attracted towards the surface (positively adsorbed), whilst ions of opposite charge to the surface (co-ions) are repelled (negatively adsorbed). The distribution of ions in this region close to the surface, the so-called diffuse electrical double layer, is given by the Boltzmann equation.[1] On the basis of this simple model for a spherical particle of radius a, the potential in the solution phase at a distance r from the centre of the particle, ψ_r, is given, for small potentials ($\psi_0 < 25$ mV), by:

$$\psi_r = \psi_0 \frac{a}{r} \exp\left[\kappa(a-r)\right] \qquad (2)$$

The parameter κ is directly related to the electrolyte concentration in the bulk phase by the relationship:

$$\kappa^2 = 2n_0 v^2 e^2 / \varepsilon_r \varepsilon_0 kT \qquad (3)$$

where n_0 is the number of ions of each type per unit volume of the bulk phase, v is the magnitude of the valency of the ions, assuming a symmetrical electrolyte, ε_r is the relative permittivity of the solution phase and ε_0 that of free space, k is the Boltzmann constant and T is absolute temperature.

ψ_0 is related to σ_0 by the expression:

$$\sigma_0 = \varepsilon_r \varepsilon_0 \psi_0 (1 + \kappa a) / a \qquad (4)$$

The quantity, κ, which has dimensions of reciprocal length, is important in that it moderates the fall-off of potential with distance from the surface, i.e. when $(r - a) = 1/\kappa$, then $\psi_{r-a} = \psi_0/\exp$. Hence at low electrolyte concentrations, for example, 10^{-5} mol dm^{-3}, $1:1$ electrolyte, $1/\kappa \approx 100$ nm and at $0 \cdot 1$ mol dm^{-3}, $1/\kappa \approx 1$ nm. The effect of the electric field extends from the particle surface a distance of the order of $2/\kappa$, e.g. in 10^{-5} mol dm^{-3} electrolyte the fields of two particles overlap when their surfaces are separated by a distance of the order of 400 nm, a long-range interaction, whereas in $0 \cdot 1$ mol dm^{-3} electrolyte the range of the interaction is two orders of magnitude smaller.

A more sophisticated model of the electrical double layer includes an allowance for some counterions adsorbing in a layer adjacent to the charged surface and at a distance d from it to form the so-called Stern layer.[1,2] The potential distribution on this basis is given in Fig. 2. There is now a linear drop between the surface and the inner layer followed by an exponential decrease with distance into the solution phase. The potential ψ_d is not easy to determine, but a quantity closely approaching it in magnitude, the zeta potential, ζ, can be determined from electrokinetic experiments.[3] One of the most direct techniques for latex particles is to determine the electrophoretic mobility, u, by a microelectrophoresis or moving boundary experiment.[4] The mobility is then related to ζ potential by the expression:[5]

$$u = \frac{\varepsilon_r \varepsilon_0 \zeta}{\eta} f(\kappa a, \zeta) \qquad (5)$$

where η is the viscosity of the dispersion medium and $f(\kappa a, \zeta)$ is a numerical factor which can be obtained from the literature.[6,7]

Figure 3 shows the electrophoretic mobility as a function of pH for a

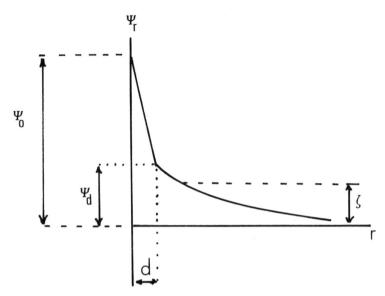

Fig. 2. Fall-off of electrostatic potential with distance from a charged latex surface.

poly(styrene) latex particle in 5×10^{-4} mol dm^{-3} sodium chloride solution. The mobility in the pH region 2–3·5 appears to be due to sulphate groups on the surface and the increase in mobility in the pH region 4–7 a consequence of the ionisation of carboxylic acid groups on the surface.[8]

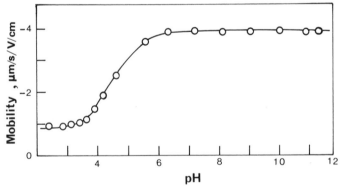

Fig. 3. Mobility against pH for poly(styrene) particles (persulphate-initiated) in 5×10^{-4} mol dm^{-3} sodium chloride solution. Particle radius = 132 nm.

2.2. Electrostatic Interaction between Two Spherical Particles

As mentioned earlier, the range of the electrostatic field around a spherical particle is of the order of $2/\kappa$. Consequently, when two particles approach to a surface–surface separation of the order of $4/\kappa$ or less overlap of the fields occurs. The overlap of the counterion clouds leads to repulsion between ions of the same charge, and hence to a repulsion between the particles. The strength of the repulsion increases as the surface–surface separation decreases. If the latter distance is taken as h (see Fig. 4) then this can be expressed quantitatively in the form:

$$V_R = 2\pi\varepsilon_r\varepsilon_0\psi_s^2 a \ln[1 + \exp(-\kappa h)] \tag{6}$$

for $\kappa a > 10$ and

$$V_R = 4\pi\varepsilon_r\varepsilon_0\psi_s^2 \exp(-\kappa h)/(h + 2a) \tag{7}$$

for $\kappa a < 3$. For intermediate κa values, V_R can be obtained either by interpolation or from an expression given by Reerink and Overbeek,[9] namely:

$$V_R = 4\cdot36 \times 10^{20}\varepsilon_r\varepsilon_0(kT)^2 a\gamma^2 \exp(-\kappa h)/v^2 \tag{8}$$

with

$$\gamma = [\exp(ve\psi_s/2kT) - 1]/[\exp(ve\psi_s/2kT) + 1]$$

V_R in these expressions is the potential energy of interaction for two spherical particles of radius a with ψ_s as the diffuse layer potential. ψ_d should be used for computations if available. In practice it often is not and the measured ζ potential is used as a good approximation. It

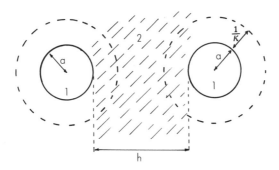

Fig. 4. Interaction between two spherical particles, radius a, of material 1 in a liquid medium 2. $1/\kappa$ = double layer parameter; h = intersurface separation.

should be noted that the exponential term arises directly from the fall-off of potential with distance (eqn (2)) and is electrolyte-dependent.

2.3. Attractive Interaction between Two Spherical Particles

The potential energy of attraction between two spherical particles of material 1 in a liquid medium of material 2 was given by Hamaker[10] in the form:

$$V_A = -\frac{A}{12}\left[\frac{1}{x^2+2x}+\frac{1}{x^2+2x+1}+2\ln\left(\frac{x^2+2x}{x^2+2x+1}\right)\right] \qquad (9)$$

with $x = h/2a$. The quantity A is called the composite Hamaker constant and for the situation cited is given by:

$$A = (\sqrt{A_{11}} - \sqrt{A_{22}})^2 \qquad (10)$$

where A_{11} is the Hamaker constant of the particles and A_{22} is that of the medium. The Hamaker constant is directly related to the nature of the material by:

$$A_{jj} = \frac{3}{4}\pi^2 h_p v_j \alpha_j^2 q_j^2 \qquad (11)$$

where h_p is Planck's constant, v_j is the dispersion frequency of the material, α_j is the static polarisability and q_j is the number of atoms or molecules per unit volume. Some typical values of A and A_{11} are given in Table 1. The value of A_{22} for water is $3 \cdot 70 \times 10^{-20}$ J.[16]

For conditions such that $x \ll 1$ a useful approximation for eqn (9) is

TABLE 1
Hamaker Constants for Various Polymers

Material	A_{11} $(10^{-20}\,J)$	A $(10^{-21}\,J)^a$	Reference
Poly(vinyl acetate)	8·84	5·50	15, 16
Poly(vinyl chloride)[b]	7·78	13·0	16
Poly(methyl methacrylate)[b]	7·11	10·5	16
Poly(styrene)[b]	6·58	9·50	16
Poly(isoprene)[b]	5·99	7·43	16
Poly(tetrafluoroethylene)[b]	3·80	3·33	16

[a] The A values for polymers lie in the range 3×10^{-21} to 13×10^{-21} J, or ca 1–3kT. The values, with the exception of PTFE, follow the order of the densities.
[b] From Lifshitz calculations for $h = 0$.

obtained as:

$$V_A = -Aa/12h \qquad (12)$$

When the value of h increases beyond ca 100 nm, a weakening of the attraction occurs due to the retardation effect. Equation (12) rewritten to allow for this has the form:[11]

$$V_A = -(Aa/h)(2 \cdot 45p - 2 \cdot 17/180p^2 + 0 \cdot 59/420p^3) \qquad (13)$$

where $p = 2\pi h/\lambda_0$. λ_0 is the dispersion wavelength and is usually taken as 100 nm. Equation (13) is a reasonable approximation for $a \gg h$ and $0 \cdot 5 < p < \infty$.

In the Hamaker theory of attraction forces between particles it is assumed that all the resonance effects occur in the ultra-violet region of the spectrum and that the whole contribution is given by v_j (eqn (11)). In some cases, however, contributions can also arise from other frequencies, for example, in the infra-red or microwave regions. An alternative approach which implicitly includes all these contributions and the effect of retardation has been given by Dzyaloshinskii et al.[12] based on quantum electrodynamics. In this treatment the constant becomes a function of separation distance and temperature, i.e. $A(h, T)$. For many polymer systems the ultra-violet dispersion contributions are dominant and $A(h = 0, T)$ is very close to the Hamaker A. For a more detailed account of this specialised topic the reader is referred to the book by Mahanty and Ninham[13] and the review by Richmond.[14]

2.4. The Total Interaction between Two Spherical Particles

The basis of current theories of colloid stability[1,17] is to consider the interaction between two particles in terms of the electrostatic repulsion, V_R, and the van der Waals attraction V_A, giving:

$$V_T = V_R + V_A \qquad (14)$$

However, as the particles approach very closely, molecular orbital overlap can occur leading to a very close-range repulsion known as the Born repulsion denoted by V_B. A more precise treatment includes this term giving:

$$V_T = V_R + V_A + V_B \qquad (15)$$

From these quantities, which are all $f(h)$, a potential energy curve of V_T against h is obtained having the form shown in Fig. 5. The curve

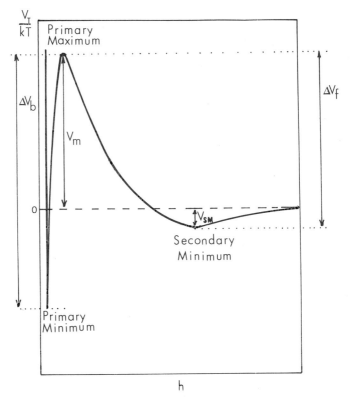

Fig. 5. Schematic illustration of a V_T/kT against h curve to illustrate the main features used in discussing colloid stability. ΔV_f = energy barrier to coagulation; ΔV_b = energy barrier to peptisation; V_m = height of the primary maximum; V_{SM} = depth of the secondary minimum.

exhibits a number of characteristic features. At short distances, a deep minimum in potential energy occurs, which is termed the *primary minimum*. Its position determines the distance of closest approach between the particle surfaces. At intermediate distances, the electrostatic repulsion makes the largest contribution and hence a maximum occurs of magnitude V_m; this is termed the *primary maximum*. At larger distances, the exponential decay of the electrical double layer causes this term to fall-off more rapidly than the inverse power law of the attractive term and another minimum occurs in the curve, of depth V_{SM}; this is termed the *secondary minimum*.

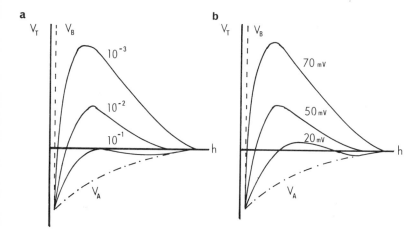

Fig. 6. Schematic potential energy diagrams: (a) influence of $1:1$ electrolyte concentration mol dm^{-3} at constant surface potential (70 mV); (b) influence of surface potential at constant electrolyte concentration (10^{-3} mol dm^{-3}).

From the various equations given we can examine the influence of electrolyte concentration and diffuse layer potential on the form of the potential energy of interaction. Two schematic examples are given in Fig. 6. It is immediately clear from Fig. 6(a) that if salt is added to the system, at constant ψ_s, the maximum is depressed until at *ca* $0\cdot1$ mol dm^{-3} the situation occurs that $V_m = 0$, or, in other words, since the particles no longer have to surmount a kinetic energy barrier, they can go into a deep energy minimum and come into close contact. Similarly, lowering the surface potential at constant electrolyte concentration leads to a potential, ψ_s, for which $V_m = 0$.

It becomes apparent from calculations of this sort that when the magnitude of V_m is $>10kT$ conditions are favourable for the formation of a stable colloidal dispersion. Moreover, the form of the potential energy curve obtained by this approach indicates that the dispersion has a higher total free energy than the corresponding amount of solid polymer. The depth of the primary minimum is equal to twice the dispersion surface free energy of the polymer.[18] The stability of the dispersion is kinetic in origin as a consequence of the large activation energy ΔV_f which gives the forward transition into the primary minimum a low probability of occurrence. It also becomes apparent that, because the activation energy for the backward reaction, ΔV_b, is much larger than ΔV_f (see Fig. 5), spontaneous redispersion is unlikely to occur and mechanical work is needed to redisperse the aggregated particles. Moreover, time has to be considered in this context since prolonged contact of particles in the primary minimum can lead to welding of the particles as a consequence of interdiffusion of the polymer chains.

2.5. The Critical Coagulation Concentration

From the previous section it becomes apparent that when V_m is substantial, $>10kT$, the probability of transition into the primary minimum is small. However, when $V_T = V_m = 0$ transition into the primary minimum can easily occur. Hence, we can define conditions for the onset of instability of an electrostatically stabilised colloidal dispersion as:[1]

$$V_T = 0 \quad \text{and} \quad \frac{\partial V_T}{\partial h} = 0 \tag{16}$$

From eqns (8) and (12) we therefore find the κ values given by these conditions as:

$$\kappa_{crit} = 2 \cdot 04 \times 10^{-5} \gamma^2 / A v^2 \tag{17}$$

or converting into molar concentration units, via eqn (3), we obtain:

$$C_{crit} = \frac{3 \cdot 86 \times 10^{-25} \gamma^4}{A^2 v^6} \text{ mol dm}^{-3} \tag{18}$$

where C_{crit} is the critical coagulation concentration, henceforth denoted as the ccc. It should be noted from eqn (18) that the ccc depends inversely on the square of the composite Hamaker constant and the sixth power of the valency of the counterions. For high potentials also, $\psi_s \gtrsim 150 \text{ mV}$, $\gamma \to 1 \cdot 0$.

For low surface potentials, $\psi_s < 25 \text{ mV}$, a further simplification can be made and with some justification we can put $\psi_s \equiv \psi_d \equiv \zeta$, and write:

$$C_{crit} \equiv ccc = \frac{8 \cdot 82 \times 10^{-19} \zeta^4}{A^2 v^2} \text{ mol dm}^{-3} \tag{19}$$

We observe from eqn (19) that if the ccc and ζ are measured an estimate of A can be made. In addition we note that this equation contains an inverse square dependence on the counterion valency, so that for univalent, divalent and trivalent counterions we obtain:

$$C_1 : C_2 : C_3 = \left(\frac{1}{1}\right)^2 : \left(\frac{1}{2}\right)^2 : \left(\frac{1}{3}\right)^2$$

$$= 100 : 25 : 11$$

This approach due to Derjaguin and Landau[17] and Verwey and Overbeek,[1] the so-called DLVO theory, has thus enabled a theoretical significance to be given to the counterion valency sequence in coagulation experiments that was observed many years earlier by Schulze[19] and Hardy.[20] Although expressions of this type are very useful as a qualitative predictive tool, the possible implication that there is a simple rule applicable to all systems must be treated with considerable caution. It must be remembered that coagulation processes do involve both kinetic and specific ion effects. The latter occur as a consequence of ion exchange in the inner part of the double layer. Hence, ions of the same valency are frequently found to form sequences which reflect their chemical properties, e.g. extent of hydration, polarisability and interaction with surface groups. A good example is found with the alkali metal ions where the concentration required to produce coagulation is frequently in the order: $Li^+ > Na^+ > K^+ > Rb^+ > Cs^+$, the lyotropic series.[3]

2.6. Coagulation as a Kinetic Process

In a fundamental paper von Smoluchowski[21] presented a theoretical model for the kinetics of the coagulation process. He showed that in the initial stages of coagulation the rate of disappearance of the primary particles, i.e. those present as single particles in the original dispersion, could be written as:

$$-dN/dt = kN_0^2 \tag{20}$$

where N_0 is number of primary particles per unit volume present initially and k is a rate constant. For rapid coagulation, i.e. coagulation in the absence of an energy barrier, the process is diffusion-controlled and $k = k_0 = 8\pi DR$ where D is the diffusion coefficient of a single particle and R the collision radius of the particle.

In subsequent analyses it was shown that if diffusion in the presence of an energy barrier is considered[22,23] then the initial rate of disappearance of the primary particles could be written as:

$$-dN/dt = k_0 N_0^2 / W \tag{21}$$

where W, termed the stability ratio, is related to V_T by:

$$W = 2a \int_0^\infty \frac{\exp(V_T/kT)\,dh}{(h+2a)^2} \tag{22}$$

In the absence of an energy barrier, i.e. setting $V_T = 0$, we find $W = 1$ and the equation for rapid coagulation is obtained. In the presence of an energy barrier V_T becomes positive and W becomes greater than unity and, clearly for these conditions, the rate of coagulation is slowed down; hence slow coagulation occurs. In practice at intermediate electrolyte concentrations and medium potentials ($\psi_s \sim 50$ mV) W can attain values of the order of 10^7 so that coagulation is imperceptible on a reasonable time-scale. The approach emphasises the kinetic nature of the stability of lyophobic colloids.

Measurements of the rate of coagulation of latex particles can be carried out by a number of techniques, two of those most commonly used being particle counting and light scattering.[24] These give values for the rate constant k and since a well-marked transition usually occurs between the slow and rapid coagulation regions the assumption is usually made that for rapid coagulation $W = 1$, whence for this region $k = k_0$, and thus in the slow coagulation region W can be obtained from the ratio k_0/k. A typical example of the type of experimental data obtained is shown in Fig. 7 plotted in the form $\log W$ against $\log C_e$, where C_e is the concentration of electrolyte in the dispersion after mixing has occurred at zero time. As can be seen, the transition between slow coagulation and rapid coagulation is clearly marked and a well-defined electrolyte concentration can be obtained from the graph at this point. This electrolyte concentration gives a value of the ccc.

Reerink and Overbeek[9] showed by an approximate treatment based on eqns (8) and (12) that for interaction at constant surface potential the gradient of the curve just before the ccc was given by:

$$d \log W/d \log C_e = -2 \cdot 06 \times 10^7 a\gamma^2/v^2 \tag{23}$$

suggesting that the slope should be directly proportional to the radius and inversely proportional to the square of the valency of the coagulating counterion. A number of experimental studies have been made that do not seem to confirm these predictions and even refinements of the kinetic treatment to allow for hydrodynamic interactions between the particles[25,26] do not seem to have removed the discrepancy be-

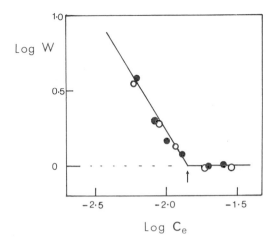

Fig. 7. Log W against log C_e curve for a poly(styrene) latex ($a = 0.21 \mu$m) in barium nitrate solutions.[24] ●, results obtained from light-scattering measurements; ○, results obtained using a particle counting technique; vertical arrow, ccc value.

tween experiment and theory. Probably further refinements are required and the assumption of constant surface potential must be questioned.

2.7. The Experimental Determination of ccc Values

From an experimental viewpoint, however, there is no doubt that the ccc is a very important quantity to know for a polymer latex, since it represents the electrolyte concentration at which complete loss of stability occurs. Experimentally, the ccc can be obtained by a variety of methods and the use of light scattering and particle counting has already been mentioned. Possibly the simplest method of all is visual observation in test-tubes containing the same concentration of latex and different concentrations of electrolyte. A slightly more elaborate version of this method is to use a simple spectrophotometer to measure the optical density of the dispersion at specific time intervals after addition of electrolyte; the most convenient time period for the comparison of data has to be determined by preliminary experiments and is often of the order of 1 h (see, for example, Ref. 27).

A time dependence of the ccc is to be anticipated since the phenomenon of coagulation is a kinetic process. There is no doubt that

TABLE 2
ccc Values for Various Polymer Latices

Latex	Counterion	ccc $(mmol\,dm^{-3})$	Reference
Poly(styrene)	H^+	1·3	28
(Carboxyl surface)	Na^+	160	29
	Ba^{2+}	14·3	24
	La^{3+} (pH 4·6)	0·3	28
Poly(styrene)	Cl^-	150	30
(Amidine surface)	Br^-	90	30
	I^-	43	30
Divinylbenzene	Na^+	160–560	31
Styrene–butadiene	Na^+	200	32
	K^+	320	32
	Mg^{2+}	6	32
	Ba^{2+}	6	32
	La^{3+} (pH 3)	0·5	32
Poly(vinyl chloride)	Na^+	50–200	33
	Mg^{2+}	2–10	33

the most precisely defined value is that obtained from the $\log W$ against $\log C_e$ curve since this is always defined in the limit $t \to 0$ if the initial rates of coagulation are used. On this basis we can anticipate that there will be some variations in the ccc values obtained by different workers according to the method that they use.

The ccc values for a number of polymer latices have been determined and some typical values are reported in Table 2. The trends observed are qualitatively in agreement with those expected from the theoretical approach for particles with smooth surfaces, with ψ_s, everywhere the same, using simple electrolytes, i.e. those which do not interact chemically with water to form new ionic species. These values should only be used for qualitative guidance since, in addition to the factors already mentioned, there can be variations of the ccc with particle size, type and density of surface groupings, and the presence or absence of stabilising materials such as surfactants. In practice it is advisable to determine the actual value for a particular latex system.

2.8. The Effects of Inorganic Ions that Interact with Water
So far, the assumption has been made that the ions used in the coagulation experiments do not interact with water. In a number of cases, however, the ions do react with water under certain pH conditions to form hydrolysed species. For example, in the case of

aluminium, the Al^{3+} ion exists at pH values below about 3·3 as the hexa-aquo ion, with six water molecules in the octahedral coordinate positions. As the pH is slowly increased, reaction occurs with water to form a sequence of species. The chemistry involved in these reactions is somewhat complex and has not been fully resolved but, in general, species of the type $Al_x(OH)_y^{n+}$ are formed. Polymeric species of this type, which are soluble in water, can adsorb strongly onto negatively charged particles and reduce the effective surface potential on the particle to zero. As anticipated earlier (Fig. 6(b)) this leads immediately to coagulation of the system. At higher concentrations of the aluminium species, superequivalent adsorption can take place, thus conferring a positive charge on the particle and leading to restabilisation of the dispersion as one containing positively charged rather than the original negatively charged particles. The additional possibility occurs that positively charged polymeric species can adsorb on to more than one particle.

The basic pattern of the coagulation of polymer latices with aluminium salts has been clearly demonstrated by the work of Matijević and co-workers[32,34,35] using styrene–butadiene, poly(vinyl chloride)[35] and poly(tetrafluoroethylene)[36] latices. The results obtained by Force and Matijević[34] for the coagulation of styrene–butadiene latices using aluminium nitrate are shown in Fig. 8. From these it can be seen that up to a pH of ~3·4 the ccc remains constant at $5 \times 10^{-4} \, \text{mol dm}^{-3}$ and then decreases between pH 3·4 and pH 4·8 to reach a constant value of $\sim 2·5 \times 10^{-6} \, \text{mol dm}^{-3}$ between pH 4·8 and pH 6·0. The region of restabilisation as positively charged particles is also clearly delineated on this type of plot.

2.9. Secondary Minimum Effects

One of the pronounced features of the V_T against h curve shown in Fig. 5 is the presence of a secondary minimum. This feature is not very pronounced for small particle sizes but becomes more distinctive and deeper as the particle size increases particularly as the electrolyte concentration is also increased. These trends are illustrated in the potential energy diagrams shown in Fig. 9. The form of the energy curve indicates the possibility that, over this range of electrolyte concentration, once a particle enters a secondary minimum it will have a long residence time there and remain separated from the second spherical particle by distances of the order of 6–10 nm. However, there remains a substantial primary maximum in potential energy to be

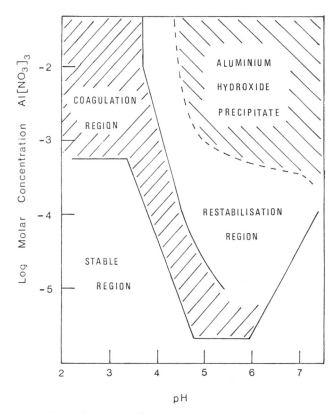

Fig. 8. Log [Al(NO₃)₃] (mol dm⁻³) against pH showing the positions of the coagulation domains for a styrene–butadiene latex. Curves constructed from the data of Force and Matijević.[34]

overcome before the particles can come into contact or enter the primary minimum. Association in the latter state clearly corresponds to a condition where the particles come into close contact, providing the possibility that with subsequent thermal diffusion they will fuse together. Under these conditions, therefore, one would expect the units formed to be hard, compact and essentially non-reversible, and there is compelling logic to term this state *coagulation*. On the other hand, when association occurs in a secondary minimum, the particles remain separated by a liquid film so that contact becomes unlikely, and leaves the possibility that by decreasing the salt concentration the particles can be redispersed. This state is usually termed *flocculation* and in a

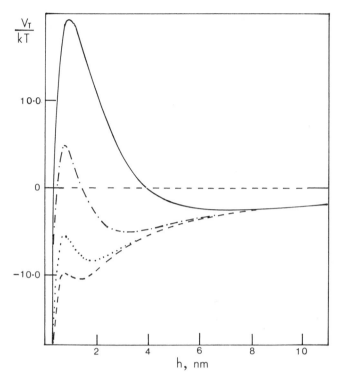

Fig. 9. V_T/kT against h for spherical particles of radius $1·62\,\mu$m at various concentrations of a $1:1$ electrolyte. ——, $0·05\,$mol dm^{-3}; —·—, $0·15\,$mol dm^{-3}; $\cdots\cdot$, $0·3\,$mol dm^{-3}; $---$, $0·4\,$mol dm^{-3}; $\psi_s = 25\,$mV; $A = 7 \times 10^{-21}\,$J at a temperature of $298·2\,$K.

number of cases it is useful to distinguish between the states of coagulation and flocculation, particularly when the systems are treated with single electrolytes so that by the continued addition of electrolyte a transition from the flocculated to the coagulated state can occur.

On a kinetic basis the presence of a secondary minimum in the energy against distance curve should lead to a steady-state condition in which the rate of particles entering the secondary minimum to form associated units should be balanced by their rate of return to single particles. Hence, assuming that doublets form the dominant associated units, the rate of disappearance of single particles can be written as:

$$-\mathrm{d}N/\mathrm{d}t = k_1 N_1^2 - k_2 N_2 \qquad (24)$$

where k_1 is the rate constant for entry into the secondary minimum, k_2 is the rate constant for exit from the secondary minimum, N_1 is the number of single particles, and N_2 is the number of doublets. For a steady-state condition we have:

$$-dN/dt = 0 \qquad (25)$$

and consequently the number of doublets is given by:

$$N_2 = k_1 N_1^2 / k_2 \qquad (26)$$

From these arguments it can be anticipated that once a steady-state condition is achieved the percentage of single particles can become constant. With a further increase in electrolyte concentration, however, a deepening of the secondary minimum can occur and therefore more particles reside in secondary minima. The experimental studies of Cornell et al.[37] appear to confirm that this is so. Moreover, these authors using poly(styrene) latices, with particles having a diameter of 2μm, and high-speed photographic techniques were able to show that the particles in an associated unit are quite mobile. It was observed that as well as some particles leaving the aggregated unit as single particles and returning to the disperse phase there was a continued rearrangement of the particles in a floccule; this was also observed at salt concentrations well above the ccc. These observations clearly support the contention that association can occur in a secondary minimum and that in this situation a liquid film is maintained between the particles.

2.10. The Effects of Organic Ions such as Surface Active Agents

If it is assumed that the latex particle has a surface which is free of adsorbed materials, that is it has been extensively dialysed or treated with mixed-bed ion-exchange resin, then the surface for an anionic particle will be composed of negatively charged sites and hydrophobic patches. Adsorption of surface active molecules can then be envisaged as shown schematically in Fig. 10.

2.10.1. Anionic Surface Active Agents

In the case where the surface active agent head group has the same sign of charge as the particle, adsorption occurs on the hydrophobic sites of the particle leaving the anionic head group exposed to the solution phase and hence increasing the overall surface charge of the particle.

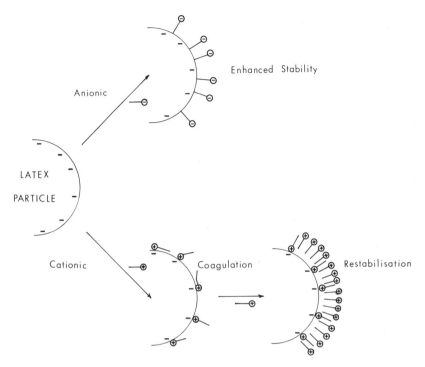

Fig. 10. Schematic diagram to illustrate the adsorption of anionic and cationic surface active agents on an anionic latex particle.

On basically hydrophobic substrates such as poly(styrene), it is well established that even on the negatively charged particles there is adsorption of surfactant anions via the hydrocarbon chains. This is demonstrated in the work of Kayes,[38] who found a substantial increase in the electrophoretic mobility of poly(styrene) latices with increase in the concentration of dodecyl sulphate in the system, and in the work of Cebula *et al.*[39] and Harris *et al.*[40] on the adsorption of dodecanoate ions on poly(styrene) latex particles.

In the case of poly(tetrafluoroethylene) (PTFE) latices, however, it has been found that, despite the fact that both fluorocarbon and hydrocarbon chains are hydrophobic, dodecanoate ions are only very weakly adsorbed on to a PTFE surface and do not even form a close-packed monolayer. On the other hand, perfluoro-octanoate ions form a monolayer at relatively low concentrations[41] and enhance the stability of the latex particles. Compatibility of the polymer particle surface and the surface active agent is therefore an important factor if the surface active agent is added to enhance the colloid stability of the system.

2.10.2. Cationic Surface Active Agents

The schematic diagram given in Fig. 10 shows that with a surface active agent having a cationic head group the first stage of adsorption is via the positive charge in order to neutralise the charge on the particle surface. At this stage ψ_s is reduced to zero, as is V_R. In this region on a hydrocarbon surface it is probable that the hydrocarbon tails lie flat on the surface. Some data obtained for the coagulation of poly(styrene) latices by a series of alkyl trimethylammonium halides are shown in Fig. 11 in the form of curves of log W against log C_e; for comparison results are also included for a simple 1:1 electrolyte, potassium bromide.[42,43] It can be seen from these data that the range of concentrations over which coagulation occurs is very narrow and that the ccc is strongly dependent on the chain length of the hydrocarbon tail of the surface molecule. Studies of the electrophoretic mobility of the particles confirm that at the ccc the particle mobility becomes zero.[44]

Once the negative charges on the particle surface have been neutralised, further adsorption of the surface occurs via the tail onto the hydrophobic patches of the surface and also by association of the hydrocarbon chains. Detailed adsorption studies on poly(styrene) latices have been reported.[44] The additional adsorption provides a positive charge to the particles and restabilisation occurs. The sharpness of this phenomenon is clearly illustrated by the experimental data given in Fig. 11. Theoretical predictions confirm that this type of behaviour would be expected on the basis of the model of the electrical double layer shown in Fig. 2.[45,46] With much higher additions of surface active agent the total electrolyte concentration increases appreciably thus reducing $1/\kappa$ and V_R, and giving a second coagulation region due to double layer compression. Once the second layer of adsorbed material is formed with the ionic head groups exposed to the solution the coagula formed are wetted and sink if the density of the particle is greater than that of the medium. When the hydrocarbon chains are exposed to the solution phase, however, as in the region of the first coagulation zone, the particles are dewetted and flotation is frequently observed.[42]

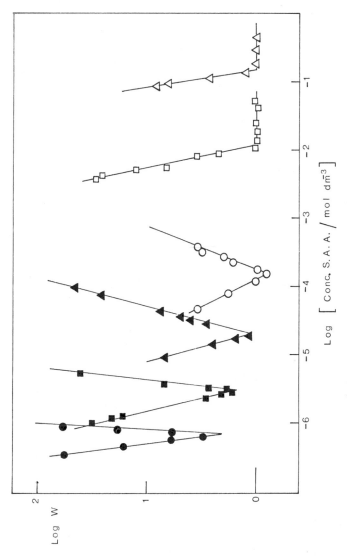

Fig. 11. Log W against the log of the concentration of alkyl trimethyl ammonium bromides of various chain lengths: ●, C_{16}; ◼, C_{12}; ▲, C_{10}; ○, C_8; ☐, C_4. For comparison, data for potassium bromide is included: △. Radius of poly(styrene) latex particles = 48 nm. (Reproduced by permission American Chemical Society.)

As can be seen from Fig. 11 with poly(styrene) latices and cationic surface active agents a systematic shift of the ccc occurs with increase in chain length indicating a typical Traube's rule effect. In the coagulation of PTFE latices by cationic hydrocarbon surface active agents, however, a different behaviour is observed in that only very small differences are observed in the ccc with variation of chain length.[46] Again it appears to demonstrate the lack of affinity of hydrocarbon chains for fluorocarbon surfaces and suggests that in this case the chains do not lie on the PTFE surfaces.

2.11. Mixed Electrolytes

In many cases, mixed electrolytes are added to latex systems either by design or adventitiously. The manner in which mixed electrolytes can cause coagulation is best illustrated as shown in Fig. 12. The axes of this figure are plotted as a percentage, so that the abscissa is the ccc value of salt 1 expressed as a percentage of its ccc value in the absence of the second salt; salt 2 is similarly expressed as a percentage of its ccc value in the absence of the first salt. The simplest possible case is *additivity* in which case the line joining the two single salt values is a straight line. When the plot is a curved line convex to both the abscissa and the ordinate the effect is called *superadditivity*. The phenomenon of electrolyte *antagonism* gives an even more pronounced convexity

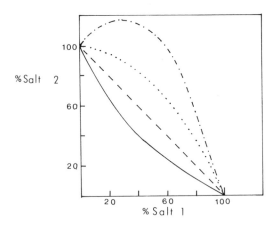

Fig. 12. Various effects obtained using mixed electrolyte systems as coagulating agents. −·−·, antagonism; ····, superadditivity; −−−, additivity; ⸺, synergism.

and the gradient at the point of intersection with the ordinate is distinctly positive. The fourth case shown in Fig. 12, when the curve is concave towards the axes, is known as *synergism*. With well-dialysed poly(styrene) latices at pH 8·5 in the presence of sodium nitrate (a 1:1 electrolyte) and magnesium sulphate (a 2:2 electrolyte) synergism was observed over the whole of the concentration range.[43,47] Moreover, the theoretical treatment of this phenomenon by Levine and Bell[48] also predicts synergism on the basis of discrete-ion effects.

2.12. Heterocoagulation

When latices of opposite charge are mixed, coagulation can occur and this phenomenon is termed *heterocoagulation*. The particles can be composed either of the same material, e.g. poly(styrene), or of different materials. In addition, small particles can also heterocoagulate with large particles if the particles have the same sign of charge and the potentials are in the right range. For particles of different radii, a_1 and a_2, and different surface potentials Hogg *et al.*[49] have given the expression for the potential energy of repulsion as:

$$V_R = \frac{\pi \varepsilon_r \varepsilon_0 a_1 a_2}{(a_1 + a_2)}$$
$$\times \left\{ 2\psi_1 \psi_2 \ln \left[\frac{1 + \exp(-\kappa h)}{1 - \exp(-\kappa h)} \right] + (\psi_1^2 + \psi_2^2) \ln \left[1 - \exp(-2\kappa h) \right] \right\} \quad (27)$$

For the van der Waals attraction an approximate expression, valid for $2a_1 \gg h$, is given by:

$$V_A = \frac{A a_1 a_2}{6(a_1 + a_2)h} \quad (28)$$

The influence of electrolyte concentration on the heterocoagulation of an anionic poly(styrene) latex of radius 26 nm and a cationic (amidine) poly(styrene) latex of radius 22 nm is shown in Fig. 13. The stability ratio was determined using an optical technique, i.e. the determination of the turbidity of the system as a function of time after adding the anionic latex to the cationic one in the presence of the appropriate electrolyte concentration.[50] The results are presented in the form log W against $N_-/(N_+ + N_-)$ with N_+ and N_- as the number concentrations of the cationic and anionic latices, respectively.

The structure of heterocoagula for particles of not very different size

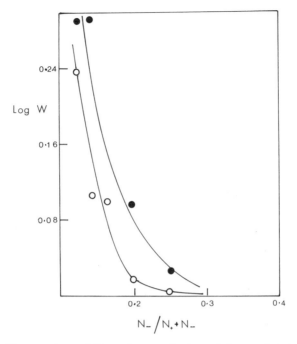

Fig. 13. Changes in stability of an anionic poly(styrene) latex (particle diameter = 52·7 nm) admixed with a cationic latex (particle diameter = 43·4 nm). N_+ and N_- = the number concentration of the cationic and anionic latices, respectively; ○, 10^{-3} mol dm^{-3}; ●, 2×10^{-3} mol dm^{-3} sodium chloride solution.

is illustrated in Fig. 14(a). However, when one particle size is very much smaller than the other complete coverage of the bigger particles by the smaller ones can occur,[43] as shown in Fig. 14(b).

An alternative example of heterocoagulation is the deposition of spherical particles on to planar surfaces. An ingenious apparatus for studying this type of coagulation was developed by Marshall and Kitchener[51] and Hull and Kitchener[52] based on the rotating disc device developed by Levich.[53] A detailed study of the heterocoagulation of poly(styrene) latex particles (diameter 418 nm) on to planar poly(styrene) surfaces in the presence of barium nitrate has been carried out by Clint et al.[54] using this technique. The onset of rapid deposition of the spheres on to the plate occurred at an electrolyte

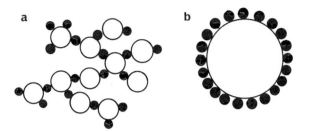

Fig. 14. Schematic illustration of (a) heterocoagulation, and (b) coating of a large particle by small particles.

concentration of $0 \cdot 02 \, \text{mol dm}^{-3}$. At this point the surface potential of the particles was $10 \, \text{mV}$ and that of the planar surface $6 \, \text{mV}$.

2.13. Surface Coagulation

In the case of polymer colloid particles, particularly those of low surface charge density, where part of the surface is hydrophobic and part hydrophilic, it is often found that coagulation can occur at electrolyte concentrations much lower than those expected for coagulation in the bulk phase, for example at the ccc values quoted in Table 2. When this phenomenon does occur it is usually connected with dewetting of the particle at the water–air interface either as a consequence of desorption of the stabilising surface active agent or because the particle surface is not homogeneous.

PTFE latices are particularly prone to surface coagulation. In part this is a consequence of their low surface energy but in addition the particles are crystalline and hence because of the polymer chain folding which occurs in the particle it appears that the ionic surface groups at the chain-ends are concentrated on some parts of the surface leaving the other regions devoid of hydrophilic groups. The latter regions have a high contact angle at the air–water interface and hence easily attach to air bubbles. At the air–water interface the particles stick together to form a coagulum which may then subsequently be redispersed by mechanical agitation.

It was found by Heller and Peters[55] that the rate of surface coagulation was given by the expression:

$$-\frac{\mathrm{d}c}{\mathrm{d}t} = \frac{K_0 S}{V}\left[\frac{c^2}{(K_1 + K_2 c)^2}\right] \tag{29}$$

where c is the concentration of colloidal particles in the bulk phase, S is the surface area of the air–water interface at a constant rate of surface renewal and V is the volume of the solution phase. K_0 is a particular rate constant, and K_1 and K_2 are constants describing the adsorption of these particles (Langmuirian) at the interface. They concluded that:

(1) Colloidal dispersions which required a relatively large amount of electrolyte in order to achieve bulk phase coagulation were not susceptible to surface coagulation.

(2) Coagulation at the surface required a low dispersion stability although the latter could be adequate to avoid coagulation occurring in the absence of a renewable liquid–air interface.

An immediate corollary to the last statement is that surface coagulation can usually be prevented by keeping latices in a container without a water–air interface.

It was also found by Heller and co-workers[55–57] that the coagula formed by surface coagulation differed from coagula formed in a bulk process. They frequently appeared to be laminar associated units, as indicated by the 'silkiness' exhibited under mild agitation.

Similar arguments to those given above can be applied to the oil–water interface and can lead to the transfer of particles from the aqueous phase to the oil phase. This can occur in polymerisation reactions under conditions where the latex particle is rapidly swollen by the monomer. This provides large oleophilic patches on the particle and a dilution of the charged areas so that particle engulfment into the monomer droplets can occur.

2.14. Peptisation

The reverse of the coagulation process is *peptisation*. The term is an old one which arises from the observation that many precipitated materials can be washed free of salt and redispersed into the colloidal state whereas this does not happen with aged material. This immediately suggests that freshly formed aggregates are not in equilibrium and that subsequently with time irreversible processes occur. Frens and Over-beek,[58] for example, have pointed out that the interpretation of peptisation phenomena with aggregated systems is not possible unless the data are obtained in experiments with a shorter time-scale than the ageing time of the aggregates. They showed that it was possible to follow the kinetics of peptisation of coagula by suddenly diluting the

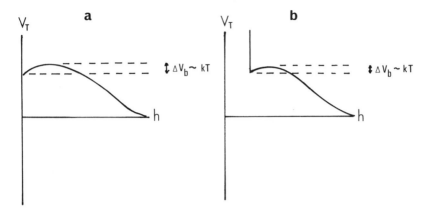

Fig. 15. Schematic potential energy curves to demonstrate favourable conditions for peptisation.

system a short time after the onset of coagulation. They concluded that peptisation was a rapid, spontaneous process and that the driving force probably arose from electrical double layer repulsion.

From Fig. 5 it is already clear that for systems with this type of potential energy barrier the transfer of particles from a coagulated, primary minimum state involves a very large activation energy, ΔV_b. Peptisation therefore is unlikely to occur in such systems. Frens[59] considers that the most favourable conditions for peptisation are those typified by the potential energy curves shown in Fig. 15. These involve a decrease in free energy in going from the associated to the free state and only a small activation energy of the order of $1kT$.

Figure 14(b) implies the presence of a short-range steric barrier on the latex particle surface such as that which might occur with 'microhairs'. The latter situation could occur by solvation of the ionic head groups such that the solvation sheath extends some way along the polymer chain (see Fig. 1). Some evidence exists for this effect in that Smitham et al.[60] have reported evidence of steric stabilisation with poly(styrene) latices which had a high carboxyl group content on the surface. Microsteric stabilisation could be an important effect. So far, it has not been fully investigated.

2.15. Coagulation—an Important Step in Emulsion Polymerisation?

One of the important factors in the process of emulsion polymerisation is the formation of polymer particles which are colloidally stable. Their stability can be ensured in the final preparation either by the addition of surface active agent[61] or in an emulsifier-free preparation by the use of an initiator which provides charged groups on the particle surface.

The progressive formation of oligomers, particle chains and ultimately particles has been considered by Fitch[62] and Goodwin et al.[63] The latter authors considered the emulsifier-free case and took the view that the first particles formed had only a small number of chains, therefore only a few end-groups were present. Consequently the particle charge was small and the system was colloidally unstable, and hence coagulation occurred until the new particles formed reached a surface charge density and radius of sufficient magnitude to give a V_R large enough to ensure colloid stability.

In order to obtain a fundamental understanding of the problem, some simple assumptions can be made. These are:

(1) The particles formed are spherical.
(2) Each polymer chain in a particle has the same molecular weight, M_c.
(3) Each polymer chain has two end-groups.
(4) All the end-groups are anchored on the particle surface.

Thus, if the latex particle has a molecular weight M_L and a density ρ_L, the number of polymer chains per particle is given by:

$$N_c = 4\pi a^3 \rho_L N_A / 3M_c = M_L / M_c \qquad (30)$$

The number of charged end-groups per particle is therefore given by:

$$N_e = 2N_c = 8\pi a^3 \rho_L N_A / 3M_c \qquad (31)$$

and the surface charge density by:

$$\sigma_s = (N_e / 4\pi a^2)e = (2\rho_L e N_A / 3M_c)a \qquad (32)$$

Thus σ_s is directly proportional to a. For spherical particles the surface potential ψ_s is given by eqn (4), an equation which holds reasonably well up to ψ_s values of 50 mV. From this we find that, taking $M_c = 150\,000$: for $a = 5$ nm, $\psi_s = 8$ mV; for $a = 10$ nm, $\psi_s \approx 20$ mV; and for $a = 22$ nm, $\psi_s \approx 50$ mV. Using a combination of eqns (7), (9) and (14) to calculate V_T as a function of h it is then possible to calculate W by numerical integration of eqn (22). Since ψ_s is known as a function of particle size, W can also be obtained as a function of r at an appropriate ionic strength.

From such calculations it becomes clear that the size of the first stable colloidal particle formed is controlled to a large extent by the ionic strength of the dispersion medium. For example, with 4×10^{-4} mol dm^{-3} $1:1$ electrolyte a W value of 10^2 is achieved with $a = 3 \cdot 7$ nm, whereas for the same W in 4×10^{-3} mol dm^{-3} salt an a value of $11 \cdot 3$ nm is required. Since, moreover, the size of the initial stable particles controls the number concentration of the latex during the diffusional growth period, for the same initiator and monomer concentrations it can be anticipated that the final particle diameter in the medium of higher ionic strength will be the larger. In experiments[64] on the preparation of poly(styrene) latices in the absence of added emulsifier, a clear trend was found in this direction. It must be remembered, however, that the total electrolyte concentration, i.e. initiator plus added salt, should be maintained well below the ccc.[63]

2.16. The Influence of a Velocity Gradient on Colloid Stability

So far, in this chapter we have considered the stability of colloidal dispersions under the influence of the earth's gravitational field. For the majority of dispersions, however, the effect of gravity is completely outweighed by thermal motion, hence the particles do not settle and we can neglect the influence of gravity; for the larger particle sizes used in *suspensions* this is not so and the gravitational term must be included. Thus for a colloidal dispersion the particle motion, translational diffusion, is controlled by solvent molecule bombardment (thermal motion $\equiv$ Brownian motion) and the interparticle forces discussed earlier. Under these conditions the flux, J (velocity $\times$ concentration) of primary particles towards another particle is given by eqn (33):

$$J = 8\pi r^2 D \frac{dN}{dr} + 8\pi r^2 \frac{DN}{kT}\left(\frac{dV_T}{dr}\right) \tag{33}$$

where N is the number concentration of the colloidal particles. Integration of this equation leads directly to the expression:

$$J = \frac{8\pi DRN_0}{R\displaystyle\int_{2a}^{\infty} \frac{\exp\left(V_T/kT\right) dr}{r^2}} \tag{34}$$

where the numerator is the flux of particles under Brownian motion alone and the denominator shows the effect of interparticle forces. The quantity R is the distance between the centres of two particles when lasting contact is formed and hence for the simplest case must be equal to $2a$. With this substitution we find immediately that the denominator is the stability ratio W defined by eqn (22), since $r = 2a + h$ and $dr = dh$.

When a colloidal dispersion is made to flow or stirred it is subjected to a velocity gradient or a rate of shear, S. Under these conditions an additional force is applied to the particles and an additional flux term has to be added[3,65] to eqn (33) giving:

$$J = 8\pi r^2 D \frac{dN}{dr} + 8\pi r^2 \frac{dN}{dT}\left(\frac{dV_T}{dr}\right) + \tfrac{4}{3}R^3 SN \qquad (35)$$

Integration of this equation leads directly to:

$$J = \frac{8\pi D N_0 R}{R \displaystyle\int_R^\infty \frac{\exp\left(V_T/kT - SR^3/6\pi Dr\right) dr}{r^2}} \qquad (36)$$

From this equation it becomes immediately apparent that the energy term due to shear has the effect of causing a reduction in the energy barrier which prevents access to the primary minimum. It is also apparent that the shear effect enters as R^3, or if the approximation is made that $R = 2a$, as $8a^3$. From eqns (8) and (12) it can be seen that the particle radius, a, enters into V_T as the first power. Hence, the shear-rate term essentially depends on a^2 and consequently it is very sensitive to particle size. For this reason small particles are rather insensitive to shearing forces, whereas large particles, $a > 0.5 \ \mu\mathrm{m}$, can often easily be flocculated by stirring or shaking. The sensitised coagulation which occurs in the presence of a velocity gradient is known as *orthokinetic flocculation*.

This effect can be assessed qualitatively in the following manner. If we take the barrier to coagulation to be V_m, then for colloid stability we require:

$$\frac{V_m}{kT} > \frac{SR^3}{6\pi D r_{max}} \qquad (37)$$

Since $D = kT/6\pi\eta a$, and as an approximation we can take $R \approx 2a$ and $r_{max} \approx 2a$, we obtain:

$$\frac{V_m}{kT} > \frac{4\eta a^3 S}{kT} \qquad (38)$$

The right-hand side can be estimated by taking $T = 20°\mathrm{C}$, $\eta = 0.01$ cP

TABLE 3
Peclet Number at Various Particle
Sizes

a (nm)	$4\eta a^3 S$ (kT)
10	10^{-4}
100	0·1
500	12·1
1000	97·0

and $S = 100 \, s^{-1}$, giving the results shown in Table 3 for the dependence on particle size.

Since a useful criterion for stability is that V_m should be greater than $10kT$, it becomes clear that dispersions containing particles with radii greater than $0·5 \, \mu m$ can be very sensitive to quite modest shear effects, particularly at high electrolyte concentrations.

Although eqn (36) is qualitatively useful as shown by de Vries,[65] it does not correctly take into account the hydrodynamic interactions which can occur between particles in flowing dispersions. Zeichner and Schowalter,[66] however, have evaluated the importance of the hydro-dynamic term by trajectory analysis and examined its effect on the stability ratio. Using a Couette viscometer they have also made experimental determinations of the effect.[67]

Following their analysis they were able to make some predictions about the behaviour of dispersions under sheared conditions and relate these to the form of the potential energy diagram. From the arguments advanced in eqn (36) it also becomes clear that when the shear energy is greater than or equal to the depth of the secondary minimum then particles can be removed from this equilibrium location and become totally dispersed. Hence a transition can occur in flow between secondary minimum association and dispersion, as shown in Fig. 16.

As already shown in the earlier sections, once the value of V_m/kT has fallen to a low value of a few kT primary minimum flocculation can occur. The analysis given above (eqn (35)) and refined by Zeichner and Schowalter[66] shows that in a velocity gradient orthokinetic floccu-lation can occur and hence the system becomes unstable at a higher surface potential than under Brownian motion alone. This is illustrated in Fig. 16 by the upward sweep of the curve between the flocculation and stable dispersion boundary. These authors also considered the question of whether deflocculation is possible under shear conditions

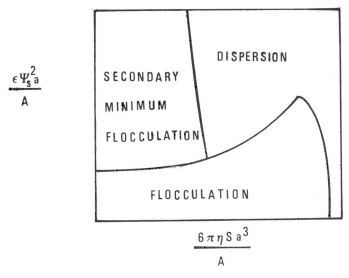

Fig. 16. Schematic illustration of the effect of shear following Zeichner and Schowalter.[66]

by examining the elongational force along the axis between two spherical particles in the direction of flow. They concluded that deflocculation could occur when the elongational force between a pair of particles became greater than the attractive force holding the particles together. This effect is illustrated by the downward curve in Fig. 16 at the high ratio of shear. The term flocculation has been used deliberately in this context since it is doubtful if particles which go into a deep primary minimum can be separated in this way. However, particles with adsorbed layers of surface active agents or macromolecules can be redispersed by shear.

Recent important work in this area has also been carried out by van der Ven and Mason[68,69] and Adler.[70]

2.17. Concentrated Dispersions
As can be seen from the preceding text, an important factor in determining the stability of an aqueous dispersion of charged particles is the range over which the electrostatic field of one particle can be felt by another. As has been shown, at very low electrolyte concentrations, the range of particle–particle electrostatic interactions can be of the

order of $1\,\mu m$, whereas at high salt concentrations it is only of the order of 1 nm or so. As a consequence of the long-range nature of electrostatic repulsion, as systems are made more concentrated, the number of particles interacting with each other increases until ultimately all the particles must be interacting with other particles. Under these conditions the particles can form a highly ordered array showing, in the size range *ca* 200 nm, pronounced optical effects such as iridescence.[71-73] The experimental techniques for examining such systems are largely those using various forms of radiation, for example light scattering, small-angle X-ray scattering and small-angle neutron scattering. These examine the microscopic structure of the system. Rheological methods, on the other hand, can provide information on the macroscopic properties of the bulk system.[160,161]

The radiation techniques in general determine a structure factor which gives the location of the particles in reciprocal space. Fourier transformation of this information gives the distribution function in real space, $g(r)$ and the radial distribution function $4\pi r^2 g(r)$.[74] The term $g(r)$ is defined by:

$$g(r) = N(r)/Np \tag{39}$$

where $N(r)$ is the number concentration of the particles at a distance r from a reference particle, and Np is the mean number concentration of particles. Typical examples are given in Fig. 17 for a poly(styrene) latex (radius 16 nm) of volume fractions 0·01 and 0·13 in $10^{-4}\,mol\,dm^{-3}$ sodium chloride solutions.[75] Curve (a) in Fig. 17 shows for the more dilute system that there is clearly an excluded volume close to the central particle. The very small peak indicates a very weak structure is formed and the dispersion will have essentially 'vapour-like' properties, i.e. the particles still have space to move around fairly freely. At a volume fraction of 0·13, however, as shown in Fig. 17(b), a very strong peak is observed in the curve of $g(r)$ against r indicating the formation of a shell of particles around the central particle at a distance of *ca* 550 Å. The second, third and fourth peaks are also visible indicating further, but more weakly ordered, shells around the central particle. This type of concentrated dispersion therefore shows very strong short-range order but rather weak long-range order; in physical terms it closely resembles the type of structures found in liquid metals.[76]

The quantity $g(r)$ can also be written in the form:

$$g(r) = \exp\left[\Phi(r)/kT\right] \tag{40}$$

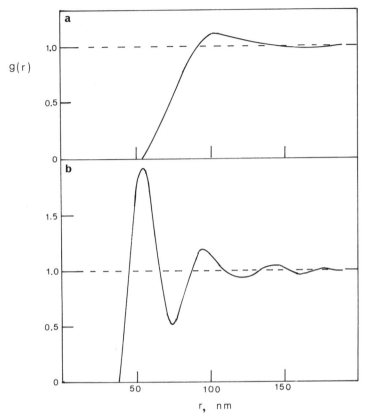

Fig. 17. Plots of $g(r)$ against r for a poly(styrene) latex ($a = 16$ nm) in 10^{-4} mol dm^{-3} sodium chloride solution. (a) Volume fraction $= 0.01$; (b) volume fraction $= 0.13$.

where $\Phi(r)$ is the potential of mean force between the particles given by:

$$\Phi(r) = V_{\mathrm{T}} + \psi(r) \tag{41}$$

V_{T} is the pair potential that has been described previously (eqn (15)) and $\psi(r)$ is a perturbation potential which allows for the many-body interactions which occur in concentrated systems. Work of this sort has already confirmed the long-range and electrolyte-dependent natures of electrostatic interactions[75–77] and the short-range nature of sterically stabilised systems.[78] The examination of concentrated dispersions[79] is still in its infancy but clearly much more work in this area can be anticipated in the near future and hopefully it should provide a useful way of predicting the behaviour of concentrated dispersions.

At high electrolyte concentrations approaching those of the critical coagulation concentration the range of the electrostatic repulsion becomes small thus diminishing the 'excluded volume' of the particle. Hence many body effects may become less important and the simple pair potential, V_T, provide a reasonable explanation of coagulation. In this context Bensley and Hunter[80] measured the ccc of anionic poly(styrene) latices, radii 177 and 356 nm, using potassium chloride and barium chloride as the coagulating electrolytes and keeping the surface potential constant so that it was independent of the latex volume fraction. They found that for K^+ and Ba^{2+} ions the ccc was independent of volume fraction over the range 0·005–0·10. They therefore concluded that for the latices examined particle volume fraction had no direct influence on the ccc. It is possible that this statement is true for non-specific ions but where specific ion effects are involved it should be examined for each system utilised.

3. THE EFFECT OF SOLUBLE POLYMERS AND NON-IONIC SURFACTANTS ON LATEX STABILITY

Soluble polymers and particularly polymers and non-ionic surfactants that adsorb at the solid–liquid interface can influence the stability of latices in a number of different ways. The addition of adsorbing species can lead to either *steric stabilisation* or *bridging flocculation* of latex particles, the outcome depending upon factors such as the amount and method of addition, and the molecular weight of the additive. Sterically stabilised latices can be reversibly flocculated by changing the affinity of the dispersion medium for the adsorbed chains, or by the addition of free, soluble polymer. Steric stabilisation and the various types of polymer flocculation are discussed in the remainder of this chapter.

3.1. Steric Stability

3.1.1. Introduction
Electrostatic stabilisation is rarely effective in non-aqueous media, it also has its limitations in aqueous media: a sensitivity to multivalent

ions has already been discussed, an equally restrictive effect in practice is the well-known but poorly understood failure of stability that often occurs at high volume fractions. The only other established means of stabilisation is steric stabilisation. Steric stabilisation can be effective in all types of media. In aqueous media sterically stabilised latices are usually much less sensitive to added electrolytes than are charge-stabilised systems. Stability can also be maintained at very high volume fractions, and this, together with the superior freeze–thaw stability and mechanical stability of sterically stabilised latices, makes steric stabilisation very useful in many technological applications.

In practice, steric stabilising agents are usually either soluble macromolecules or non-ionic surfactants; in the latter case, especially those with bulky solvatable head groups such as the alcohol ethoxylates, for example dodecylhexaoxyethylene glycol monoether $(C_{12}H_{25}(CH_2CH_2O)_6OH$—'$C_{12}E_6$'). The adsorption of such materials onto latices is considered in Chapter 4. Steric stability arises essentially from a reluctance of solvated adsorbed layers to interpenetrate as particles collide. Thus two very basic requirements for stability are the presence of a complete or entire interfacial layer so that stabilising species cannot migrate laterally out of the interaction zone between two particles, and strong adsorption so that desorption from the interaction zone cannot occur. Also, since the density of chain molecules in the adsorbed layer is rarely in practice of a level sufficient to screen the van der Waals attraction between the underlying latex particles, an additional requirement for good stability is that the adsorbed layer should have a thickness greater than the effective range of the van der Waals attraction.

3.1.2. Types of Stabilising Molecule

Soluble homopolymers can sometimes be effective as stabilisers for latices. However, they do not in general show the strong, irreversible adsorption necessary for good stability. Amphipathic molecules containing lyophobic groups that promote adsorption are thus usually more effective. Examples of amphipathic stabilisers are the alkanol and alkylphenol ethoxylate surfactants, and block and graft copolymers comprising distinct stabilising and anchoring chains (Fig. 18). The design and synthesis of amphipathic polymeric stabilisers for use in non-aqueous media have been discussed in detail by Barrett,[81] and non-ionic surfactants for use in aqueous media by Schick.[82] The ideal polymeric stabiliser would arguably be one where all of the segments

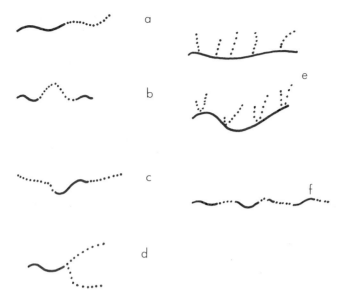

Fig. 18. Types of copolymer stabiliser: a, AB block copolymer; b, ABA block; c, BAB block; d, simple graft copolymer; e, comb-grafts; f, copolymer of ill-defined structure. ——, Anchor chain; – – –, stabiliser chain.

of the anchor component are adsorbable leaving terminally anchored solvated chains penetrating into the dispersion medium. The comb-graft type of molecule probably comes closest to this ideal, and has been found to be very effective in practice.[83] The optimum polymer composition in terms of the proportion of anchor to soluble component appears to be of the order of 1 : 1 by weight in the case of molecules for use in non-aqueous media. Copolymers and surfactants having a similar anchor–stabiliser balance are also found to be effective in aqueous environments.[83,84] The precise choice of anchoring component can be complicated by considerations such as possible competition between tendencies to undergo irreversible micellisation in solution and irreversible adsorption, and is often limited by synthetic considerations. These and other points have been discussed by Walbridge (in Ref. 81) in the context of non-aqueous systems; many of the guidelines given may nevertheless be adapted for, or extrapolated to, aqueous media. Certain of the problems associated with the synthesis of polymeric stabilisers are minimised by having stabiliser chains of modest molecular weight. For this reason the question of how large the stabilising moieties need to be in order to effect good stability is an important one. Experimental evidence relevant to this point will be discussed later in this chapter.

3.1.3. Theory of Steric Stability

In the discussion of electrostatic stabilisation attention was focused on the pairwise interaction potential V_T; this had two long-range components, V_R the electrostatic repulsion and V_A the van der Waals attraction. An additional term has to be added in order to take into account a steric contribution to stability so that now:

$$V_T = V_A + V_R + V_S \qquad (42)$$

Note that, in general, V_R has to be retained since stability could well result from a combination of steric and electrostatic repulsions. Nevertheless, V_R can often be ignored, for example where the ionic strength of the medium is high or where the dielectric constant of the medium is such that it will not support ionisation. If the surface layers are sufficiently thick, then V_A, which again is taken to be the van der Waals attraction between the underlying latex particles, can also be neglected. Thus in certain circumstances, stability can be controlled solely by the form of V_S. V_S, unlike V_R, can be either repulsive or attractive between identical particles depending upon the affinity of the dispersion medium for the adsorbed chains. In what follows, attention will be concentrated primarily on the role of V_S. The interplay of all three long-range forces will be considered explicitly later when the effect of low molecular weight non-ionic surfactants on the stability of aqueous latices is dealt with.

The theory of steric stability attempts to estimate V_S using approximate statistical–mechanical models of the surface layer. A number of detailed treatments now exist, and a comprehensive discussion of these, together with an exhaustive set of references, can be found in a review by Vincent and Whittington.[84] In many respects, the various treatments are similar in that they apply the methods used in various mean-field theories of polymer solutions to the problem of estimating the free energy of overlap of two layers of solvated chain molecules. They differ in their rigour and complexity, and in the details of the simplifying assumptions made. Nevertheless, their predictions concerning the behaviour of sterically stabilised dispersions are qualitatively very similar. The basic mechanism of steric stabilisation is thus well established, although this is not to say that many uncertainties con-

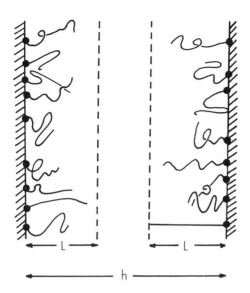

Fig. 19. Schematic illustration of two interacting surfaces bearing terminally anchored polymer chains.

cerning the detail do not remain. In this respect, the current level of understanding of steric stabilisation is rather similar to that of electrostatic stabilisation in that the theory provides a sound basis for a broad understanding rather than a fully quantitative description of stability and flocculation.

Figure 19 depicts schematically the interaction of two rigid surfaces covered with layers comprising terminally anchored flexible polymer chains of contour length L. For simplicity, the chains are assumed to be monodisperse. At separations $h < 2L$ the layers of necessity interfere with each other to some extent, and so have either to interpenetrate, or indent, or to undergo a combination of the two. Whatever the case, the segment density in the interaction zone has to increase causing a change in the local osmotic pressure. The associated change in free energy can be estimated approximately using Flory–Krigbaum theory.[85–87]

The theory of Flory and Krigbaum gives for the free energy of mixing of chain segments and solvent within a small volume element δV:[88]

$$\delta(\Delta G_{\mathrm{m}}) \simeq \frac{\mathbf{k}T}{V_1}[(\chi - 1)\phi_2 + (\tfrac{1}{2} - \chi)\phi_2^2] \tag{43}$$

where ϕ_2 is the volume fraction of segments within δV, V_1 is the volume of a solvent molecule and χ is the Flory segment–segment interaction parameter. In Flory's original treatment of polymer solutions χ was given by:

$$\chi = -2 \, \Delta\varepsilon_{12}/\mathbf{k}T \tag{44}$$

where $\Delta\varepsilon_{12}$ is a temperature-independent contact potential representing the net interaction of two segments immersed in a solvent, and 2 is a numerical factor (lattice coordination number) which need not concern us here. More generally, χ might be regarded as an empirical but experimentally measurable interaction parameter which characterises the solvency of the medium for the polymer chains. In order to account for the properties of real polymers, χ has usually to become a function of temperature, pressure and segment concentration. For simple polymers dissolved in simple solvents χ is dominated by van der Waals interactions. The net van der Waals force between two chemically identical segments is always attractive and so $\Delta\varepsilon_{12}$ is negative and χ positive in most cases. This being so, it is evident from eqn (43) that the partial demixing of chain segments and solvent that occurs in the interaction zone between two polymer-coated surfaces can cause either an increase or a decrease in free energy depending upon the magnitude of the interaction parameter. There is thus a combinatorial or *mixing* contribution to the steric interaction potential which can be either attractive or repulsive depending upon the solvency of the medium for the stabilising chains (as measured by χ). With regard to the precise mechanism of the interaction it is perhaps not entirely clear from the outset whether interpenetration or indentation, or a mixture of the two is the more likely mode of interaction. A rigorous theory would be capable of deciding this implicitly, but at present an assumption has to be made. Nevertheless, in any event, some restriction or compression of the space available to the chains has to occur at some juncture, for instance at separations $h < L$ if free interpenetration of the layers is assumed to take place. The resulting loss of configurational freedom experienced by the anchored chains makes an additional entropic or *elastic* contribution to V_S which is purely repulsive. The steric interaction thus has two components: a mixing component which may either be attractive or repulsive depending upon the value of χ, and an elastic component which is purely repulsive. The evaluation of these two contributions requires a knowledge of the segment-density distribution between the surfaces. The segment-density distribution itself, however, reflects a similar balance of short-range (mixing) and long-range (elastic) effects throughout the adsorbed layer. Thus, although it is conceptually useful to think of V_S as comprising distinct mixing and elastic contributions, these effects are coupled by virtue of their mutual dependence on and effect on the segment-density distribution. The necessary self-consistent evaluation of V_S is, however, a formidable problem which has only been attempted approximately for conditions of good or marginal solvency, that is for $\chi < 0.5$.[89-91] In this region the separation of V_S into independent mixing and elastic components turns out to be a crude albeit useful approximation and it will be employed here.

The various approximate theoretical approaches can be grouped or classified in a number of different ways. Napper[92] has distinguished between *pragmatic* theories[85,86,93,94] which employ an assumed form for the segment-density distributions in the adsorbed layers, and *ab initio* theories[87,89,91] that attempt to calculate the segment-density distributions from first principles. The *ab initio* theories may then be subdivided into classical theories[87] which use the methods described by Flory[88] to estimate the mixing and elastic contributions to V_S, and the self-consistent field theories,[89-91] mentioned above, which avoid an artificial separation of the configurational and mixing problems. It has been shown that the classical and self-consistent field theories can be made to yield very similar results, provided that a simple empirical excluded-volume correction to the segment-density distributions is incorporated in the former.[95] With the pragmatic approach, there is the problem of deciding whether the interaction between two surfaces involves free interpenetration of the surface layers[85,86,93] or 'denting'.[94,96] *Ab initio* theories implicitly assume that the surface layers interpenetrate; it is, however, possible using the pragmatic approach to explore the consequences of free interpenetration as the mode of interaction as one extreme and denting as the other. The desire to do this stems in part from concern that the time-scale of a Brownian collision may not be sufficient to allow the adsorbed chains to relax to their equilibrium configurations during the course of a particle–particle encounter. It has thus been suggested that calculations made using the denting model may be more appropriate to the problem of dispersion stability.[96] A dynamic approach incorporating both direct and hydrodynamic interactions between the chains would, however, be required in order to resolve this problem, whereas in these terms the denting model is a crude attempt to correct the direct part only. It will be seen below that the assumption of thermodynamic equilibrium appears to account for many of the observed properties of sterically stabilised latices and so this question will not be pursued, a pragmatic view is thus taken here. The denting model remains of interest nevertheless, since it has also been seen as appropriate where the adsorbed layers consist largely of irreversibly adsorbed polymer loops,[94] as might arise, for example, from the adsorption of a BAB block copolymer (in which A is the solvated chain).

Strictly speaking, all of the current theoretical models are restricted in their application to considerations of stability in good or marginal solvents, that is $\chi < 0.5$, since none is capable of reproducing the collapse of the adsorbed layers that is expected to occur in poor solvents $(\chi > 0.5)$. It can, however, be argued that the pragmatic approach is capable of providing some guidance in this region since this effect can be incorporated to some extent by making a judicious choice for the form of assumed segment-density distribution function.

In terms of their results, the various theoretical treatments differ primarily in their assessment of the relative importance of the mixing and elastic effects. The most marked difference arises between the free-interpenetration and denting type models, although the form of the initial $(h \to \infty)$ segment-density distribution also has an effect. Theoretical predictions concerning steric stability and the differences that the various possible assumptions make will be illustrated using results taken from the work of Napper and co-workers,[92–94] two very simple segment-density distributions will be employed: the uniform distribution and the exponential distribution.

The uniform distribution has the form:

$$\phi_2(2) = \bar{\phi}_2; \qquad 0 < 2 < L$$
$$= 0; \qquad 2 > L \tag{45}$$

and the exponential distribution:[94]

$$\phi_2(2) = \bar{\phi}_2 \beta L \frac{\exp(-\beta 2)}{[1 - \exp(-\beta L)]} \tag{46}$$

where 2 is the distance normal to the surface, $\bar{\phi}_2$ is the average volume fraction of segments in the adsorbed layer and β is a constant that characterises the steepness of the decay in segment density away from the surface. It might be argued that these two distributions represent useful bounds since most real segment-density distributions (see Chapter 4) are likely to have forms intermediate between these extremes. Thus the uniform distribution might be thought of as a limiting approximation to the case of short, terminally anchored tails of identical molecular weight, as may arise, for example, from the adsorption of a pure non-ionic surfactant,[85] or a comb graft of low molecular weight and polydispersity; the exponential distribution has been suggested as a simple approximation for the case of loops,[94] and so may be more appropriate for adsorbed homopolymers and BAB block copolymers (where A is the stabilising chain).

The assumption of free interpenetration gives in obvious notation:

$$V_S(h) = \Delta G_M^{FI}; \qquad\qquad L < h < 2L \qquad (47)$$

$$= \Delta G_M^{FI} + \Delta G_{EL}^{FI}; \qquad h < L \qquad (48)$$

which for spherical particles becomes:[93]

$$V_S = \frac{2\pi a \mathbf{k} T}{V_1} \bar{\phi}_2(\tfrac{1}{2} - \chi) S_M; \qquad\qquad L < h < 2L \qquad (49)$$

$$= \frac{2\pi a \mathbf{k} T}{V_1} \bar{\phi}_2(\tfrac{1}{2} - \chi) S_M^* + \frac{2\pi a \mathbf{k} T}{V_2} \bar{\phi}_2 S_{EL}; \qquad h < L \qquad (50)$$

where S_M, S_M^* and S_{EL} are geometric functions of h, L and β that have been evaluated for a number of simple distribution functions including the two of interest here.[93,94] For example, the uniform distribution gives:

$$S_M = 2(L - h/2)^2 \qquad (51)$$

$$S_M^* = L^2[3 \ln (L/h) - 2h/L - \tfrac{3}{2}] \qquad (52)$$

$$S_{EL} = L^2[\tfrac{2}{3} - \tfrac{1}{6}(h/L)^2 + h/L \ln (h/L)] \qquad (53)$$

The corresponding results for the denting model are obtained simply by replacing h by $h/2$ in equations and using these over the whole range $h < 2L$. Before the various models are compared, several general points deserve mention.

It can be seen that V_S is of order:

$$V_S \sim a L^2 \mathbf{k} T / V_1 \qquad (54)$$

or since $L \simeq V_1^{\frac{1}{3}} N$, where N is the number of segments per chain:

$$V_S/\mathbf{k}T \sim aN/V_1^{\frac{1}{3}} \qquad (55)$$

and thus since the particle radius a is typically of order $10^3 V_1^{\frac{1}{3}}$ for latices and N may be large, $V_S/\mathbf{k}T$ can clearly be large.

The mixing term in equations changes sign at $\chi = 0.5$. For a given polymer in a given solvent, χ depends primarily upon temperature and pressure, the effect of temperature on V_S being of particular importance. The term $(\frac{1}{2} - \chi)$ may be rewritten as:[88]

$$\frac{1}{2} - \chi = \Psi_2(1 - \Theta/T) \tag{56}$$

where Θ is Flory's theta temperature and Ψ_2 is the entropy of mixing parameter. In general, Ψ_2 depends upon temperature, but it can be regarded as a constant at temperatures close to Θ. The sign of $(\frac{1}{2} - \chi)$, and thus of the mixing term, clearly depends upon the sign of Ψ_2 and the value of the reduced temperature Θ/T. In the case of stabiliser chains having an upper critical solution temperature (UCST), $\Psi_2(T \simeq \Theta)$ is positive, and so the mixing term changes from being positive (repulsive) to negative (attractive) as the temperature is reduced through Θ. For chains having a lower critical solution temperature (LCST) Ψ_2 is negative in the region of $T = \Theta$ and so the converse is true and the mixing term becomes attractive as the temperature is raised through Θ.[98] With the free-interpenetration model only the mixing term contributes to V_S when $L < h < 2L$ and so V_S itself changes sign as the temperature is either raised (LCST) or lowered (UCST) through the appropriate Θ-temperature. At closer approach this is not the case, as the elastic term dominates when $h < L$. This is evident from eqns (49) and (50) which show that the mixing term is of the order $\bar{\phi}_2/V_1$ whereas the elastic term is of the order $\bar{\phi}_2/V_2$ (V_1 and V_2 will generally be similar, and $\bar{\phi}_2$ is usually very much less than 1). In poor solvents V_S is attractive when $h > L$, but repulsive at closer separations and so in the language used in the context of electrostatic stabilisation V_S can show a secondary minimum when $\chi > 0.5$. A form of flocculation, *incipient* flocculation, is thus expected to occur in sterically stabilised dispersions under conditions of poor solvency. This flocculation is also expected to be reversible, since reversing the temperature change should remove the secondary minimum and convert V_S into a repulsion at all separations.

The use of the exponential distribution in place of the uniform distribution, or the denting model in place of the free-interpenetration model, does not alter these general conclusions. The various models differ only in their predictions of the extent of heating (LCST) or cooling (UCST) past the Θ-point required to produce a discernible minimum in V_S. This point is illustrated in Table 4 which shows the calculated temperature difference $|T - \Theta|$ required to produce a

TABLE 4
Amount of Penetration Past the Θ-Point Needed to Produce a Minimum of
$-5kT$ in V_s[94]

| Stabiliser/medium | CST | Free interpenetration | | Denting Exponential |
		Uniform distribution	Exponential	
Poly(vinyl alcohol)/aqueous	LCST	<1	7	17
Poly(ethylene oxide)/aqueous	LCST	<1	15	36
Poly(acrylic acid)/aqueous	UCST	<1	11	25
Poly(styrene)/cyclohexane	UCST	<1	27	55
Poly(isobutylene)/toluene	UCST	<1	19	39

minimum $5kT$ deep in V_s. The calculations, which are due to Smitham and Napper,[94] were made using thermodynamic data (Ψ_2 and Θ) chosen for various stabiliser chains as indicated. The particle radius was taken to be 100 mm, and $\bar{\phi}_2$ and the layer thickness L were chosen so as to mimic earlier experimental work. The use of the free-interpenetration model, together with the uniform segment-density distribution function, leads to the expectation that a sensible secondary minimum should develop very close to the Θ-point. The correspondence is weaker when the exponential distribution is used and weaker still for the denting model, where elastic effects are important at all separations $H < 2L$.

The theory therefore predicts a loss of stability under conditions of poor solvency, but whether or not the transition from stability to instability should occur close to the relevant Θ-point is not entirely clear, since predictions concerning this are sensitive to the structure assumed for the adsorbed layers, to Ψ_2, and to assumptions about the precise nature of the steric interaction. Further, should the adsorbed layers be very thin (perhaps $L < 10$ nm) then the van der Waals attraction between the underlying particle would also be expected to have an influence. In pure solvents the Θ-temperature depends solely on pressure. However, in mixed solvents and in solvents containing extraneous solutes, solvent composition has also to be considered, the presence of dissolved electrolytes being of particular importance in aqueous systems. The stability of sterically stabilised latices is thus found to be sensitive to temperature, pressure and the addition of solutes to the dispersion medium, to an extent which is in accord with their effect on the solvency of the medium for the stabiliser chains.

3.1.4. Direct Measurement of Steric Interactions

Measurement of the resistance to compression of two-dimensional rafts[99] and three-dimensional arrays[100] of latex particles has been used to determine force of interaction (dV_S/dh). Such measurements are only practicable under conditions where there is a repulsion, nevertheless they have shown that a strong repulsion does exist in good and marginal solvents $(\chi < 0\cdot5)$. Direct measurements of a repulsion have also been made using polymer layers adsorbed on macroscopic surfaces.[101,102] The detection of a force of attraction is more difficult; however, recent measurements by Klein[103] have demonstrated the existence of a steric attraction between two mica cylinders separated by a solution of polystyrene in a poor solvent. Similar measurements made using chemically anchored chains or block and graft copolymers have yet to be made but would be of great interest.

3.1.5. Flocculation of Sterically Stabilised Latices

It is now well established that the solvency of the dispersion medium for the stabilising chains is one of the principal factors that influences the stability of sterically stabilised latices provided that:

(1) The surface coverage and anchoring of the stabilising chains are such that desorption or lateral migration of the stabiliser away from the interaction zone does not occur.

(2) The surface layers are of a thickness sufficient to cause the van der Waals contribution to the long-range interaction potential to be attenuated.

The first condition means in practice that good stability is normally only obtained in the plateau region of the adsorption isotherm. The second is more difficult to quantify, but it suggests that the layer thickness should be of the order of 10 nm (or more) rather than 1 nm; the effect of stabiliser molecular weight and layer thickness will, however, be discussed in detail below.

Flocculation can be caused by a change in temperature, pressure or solvent composition provided that the chains have an accessible critical point. The onset of instability normally occurs as a sharp transition from indefinite stability to rapid flocculation, the transition occurring over perhaps 1 or 2 K when temperature is the variable of interest. Because of this, the kinetics of flocculation are of secondary importance, and it is usually adequate simply to define a critical flocculation point (temperature, pressure or composition) in order to describe the behaviour of sterically stabilised systems. The existence of CFT, CFP and CFC has now been amply demonstrated.[81,92,94] The theory points to a correlation between the CFP and the appropriate Θ-point for chains in free solution of variable and somewhat uncertain exactness. This expectation is borne out by the experimental evidence, which will now be discussed.

3.1.6. Incipient Flocculation of Non-Aqueous Systems
Of the possible means of causing flocculation, a change in temperature has been the most widely studied. Latices having LCFT, UCFT and both LCFT and UCFT have now been prepared. A comparison of selected CFT and Θ-temperature data is made in Table 5. The superscript + or − in the last column indicates whether flocculation occurred under better than (+) or worse than (−) Θ-conditions. The data reveal a clear parallel between CFT and Θ on the average; there are, however, significant discrepancies in certain instances. The lack of dependence of CFT on molecular weight shown by the poly-(isobutylene)/2-methylbutane and poly(dimethylsiloxane) data suggests that here V_A had no influence on the flocculation. In spite of this, flocculation occurred under better than Θ-conditions in certain other instances, in particular the poly(styrene)/cyclopentane data would ap-

TABLE 5
Critical Flocculation Temperatures for Non-Aqueous Latices

Stabiliser	Molecular weight	Medium	Type	CFT (K)	$|Θ - CFT|$ (K)
Poly(isobutylene)[104,112]	$2·3 \times 10^4$	2-Methylbutane	UCST	325	0
	$1·4 \times 10^5$	2-Methylbutane		325	0
	$7·6 \times 10^6$	2-Methylbutane		327	$\sim 2^+$
	$7·6 \times 10^6$	2-Methylpentane		381	5^-
	$7·6 \times 10^6$	2-Methylhexane		423	3^+
	$7·6 \times 10^6$	3-Ethylpentane		463	5^-
	$7·6 \times 10^6$	Cyclopentane		455	6^+
Poly(styrene)[105]	$1·1 \times 10^5$	Cyclopentane	UCST	410	17^+
	$1·1 \times 10^5$	Cyclopentane		280	13^+
Poly(α-methylstyrene)[106]	$9·4 \times 10^3$	n-Butylchloride	LCST	254	9^-
	$9·4 \times 10^3$	n-Butylchloride		403	9^-
Poly(dimethylsiloxane)[107]	$3·2 \times 10^3$	n-Heptane/ethanol	LCST	340	~ 0
	$1·1 \times 10^4$	n-Heptane/ethanol		340	~ 0
	$2·3 \times 10^4$	n-Heptane/ethanol		341	~ 0
	$4·8 \times 10^4$	n-Heptane/ethanol		338	~ 0

TABLE 6

The Effect of Solvent Composition on the Flocculation of Non-Aqueous
Latices (taken from Ref. 81)

Stabiliser	Flocculant/solvent	CFC (% V/V)	|CFC − Θ| (% V/V)
Poly(hydroxystearic	Ethanol/n-heptane	39·5	0·5+
acid)	n-Propanol/n-heptane	61	4−
	n-Butanol/n-heptane	74	8+
Poly(lauryl	Ethanol/n-heptane	38	1−
methacrylate)	n-Propanol/n-heptane	56	4+

pear to show a real and significant penetration into good solvency
conditions. At first sight the theory offers no obvious explanation for
such an effect when the stabiliser molecular weight is so high. However, the theory assumes implicitly that the segment density in the
adsorbed layer is uniform laterally and thus strictly only applies in the
limit of high surface concentration, differences in surface density may
thus account for the differences in behaviour between various systems
and the extent to which there is a correlation between CFT and Θ.
Table 6 shows a similar comparison, but now solvent composition is
the variable of interest. Again, while a tolerable correspondence
between CFC and Θ is observed, the correspondence is not exact.

3.1.7. Incipient Flocculation of Aqueous Systems

Water-soluble polymers rarely display CST in the normal liquid range
of 273–378 K when the medium is pure water, an exception being
poly(ethylene oxide) which has an LCST just below 373 K at high
molecular weight.[82] However, the addition of low molecular weight
solutes can often produce an accessible CST. The effect of electrolytes
is of particular interest and importance.

Table 7 shows CFT data for latices stabilised by three common
water-soluble polymers. The extreme right-hand column contains notional Θ-point data for the stabilising chains which deserve some
preliminary explanation. Θ-temperatures are normally determined by
extrapolating critical temperatures obtained from phase-boundary or
light-scattering measurements to infinite molecular weight. Such measurements are tedious to perform, and very often the necessary range of
samples of different molecular weight is not available. The notional
values shown here were obtained using an empirical extrapolation
procedure of the type described by Cornet and van Ballegooijen.[113]
The use of this method, which has some theoretical basis for the binary

TABLE 7

Critical Flocculation Temperatures for Aqueous Latices

Stabiliser	CST	Molecular weight	Medium	CFT (K)	'Θ' (K)
			(mol dm^{-3} MgSO$_4$)		
Poly(ethylene oxide)[108,109]	LCST	750	0·163	316 (±2)	345
		750	0·26	308	332
		2 000	0·26	315	332
		10^4	0·39	318	315
		$9·6 \times 10^4$	0·39	316	315
		10^6	0·39	317	315
			(mol dm^{-3} HCl)		
Poly(acrylic acid)	UCST	9 800	0·2	287 (±2)	287
(unionised)[110,111]		$1·6 \times 10^4$	0·2	285	287
			(mol dm^{-3} NaCl)		
Poly(vinyl alcohol)[112]	LCST	$2·6 \times 10^4$	2	302 (±3)	300
		$5·7 \times 10^4$	2	301	300
		$2·7 \times 10^5$	2	312	300

systems for which it was developed, has, however, never been substantiated for ternary systems such as polymer/water/salt; there is thus no *a priori* reason to believe that the method works. The notional Θ-points shown here may thus differ from the true Θ-points. There is, however, a strong parallel between the CFT and notional Θ-point, and so it would appear that the latter is a useful guide to the effect of solvency on colloid stability nevertheless.

Electrostatically stabilised latices are sensitive to electrolytes containing divalent, and more generally multivalent, counterions, for example anionic latices are usually coagulated by a concentration of barium or calcium chloride which is about one-twentieth of the corresponding amount required of sodium chloride. The effect of the nature of the cation on the flocculating power of simple electrolytes for latices stabilised by poly(ethylene oxide) has been investigated by Napper.[114] The flocculation concentrations were in the order:

$$Ba^{2+} \simeq Mg^{2+} \simeq Ca^{2+} \simeq Li^{2+} > Sr^{2+} \simeq NH_4^+ > Cs^+ \simeq Na^+ \simeq K^+ \simeq Li^+$$

monovalent ions being generally more efficient flocculants than divalent ions. The effect of anions might be deduced from their effect on the notional Θ-point;[115] the following order of CFC would then be anticipated:

$$NO_3^- > Cl^- > CH_3COO^- > F^- > S_2O_3^{2-} > CO_3^{2-} \simeq SO_4^{2-} > PO_3^-$$

Notice that cations and anions have the reverse valency-dependence.

3.1.8. Polyelectrolyte Stabilisation

Early work[116] on steric stabilisation showed that adsorbed polyions could have an effect on the stability of aqueous colloids that could not be accounted for solely in terms of any enhancement of the electrostatic repulsion that their adsorption might cause. It has been shown more recently that it is possible, using synthetic polyelectrolytes, to produce latices that are stable in electrolyte solutions with ionic strengths in excess of $5 \, \text{mol dm}^{-3}$. The electrostatic repulsion between two particles is highly screened under such circumstances and so it is clear that polyelectrolytes must act as steric stabilisers in media of high ionic strength.

Figure 20 depicts a latex particle coated with a layer of solvated polyelectrolyte chains. The latter are for the moment assumed to be anionic and strongly acidic, so that the counterions are substantially dissociated. A large proportion of the counterion population will nevertheless remain within the domain of the adsorbed chains as a consequence of the high electrostatic potential in this region. The space-charge density in the outer diffuse layer, however, may be substantial and comparable to that surrounding a conventional anionic latex. Thus, in aqueous media of low ionic strength ($I \ll 10^{-2} \, \text{mol dm}^{-3}$), there will be a long-range repulsion between particles which is similar but, in view of the elasticity of the polyelectrolyte layer, a little 'softer' than the double layer repulsion between conven-

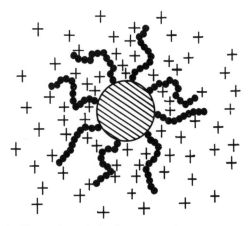

Fig. 20. Schematic illustration of the ion atmosphere surrounding a particle stabilised by anionic polyelectrolyte chains.

tional charged latices. At higher ionic strength the long-range electrostatic repulsion will, however, be screened since the Debye length κ^{-1}, which for example is only 1 nm at an ionic strength of $0 \cdot 1 \ mol \ dm^{-3}$, will not be much larger than the size of a single monomer unit making up the polyion chain. Under such circumstances it is clearly no longer useful to think of the interaction of two polyelectrolyte layers as occurring through the overlap of extended ion-atmospheres. Rather it is more appropriate to focus attention on the electrostatic repulsion between individual segments and to regard this as just one of a number of intermolecular interactions that contribute to the χ-parameter. Support for this type of approach can be gained from the observation that polyelectrolytes dissolved in concentrated salt solutions have conformational and thermodynamic properties very similar to non-ionic polymers dissolved in good solvents.

The segmental electrostatic repulsion will clearly make a negative contribution to χ and so tend to oppose segregation phenomena such as phase separation in solution and the incipient flocculation of latices. Nevertheless, polyelectrolytes having upper and lower CST are known, examples being poly(acrylic acid) and its salts (UCST), and poly-(methacrylic acid) and its salts (LCST), and latices with upper and lower and both upper and lower CFT have been prepared.[117-119]

Θ-temperatures for polyelectrolytes depend upon the concentration and type of ionic species present in solution, and in the case of weak acids and bases upon the prevailing state of acid–base equilibrium. The weak polyacid case will be considered by way of illustration and the poly(acrylic acid)/sodium polyacrylate system will be taken as a specific example. The degree of ionisation α' at any pH is related to pH and composition by:

$$\alpha' = \alpha + \frac{10^{-pH} - 10^{pH-14}}{[COOH]} \tag{57}$$

$$\alpha = [NaOH]/[COOH] \tag{58}$$

$$pH = pK_a(\alpha) + \log_{10}\left(\frac{\alpha'}{1-\alpha'}\right) \tag{59}$$

where $pK_a(\alpha)$ is the effective pK_a of the polyion at any degree of neutralisation α. α is written here as the ratio of the notional concentration of added base to total concentration of carboxyl groups, so that $\alpha = 0$ for pure poly(acrylic acid) and $\alpha = 1$ for pure sodium polyacrylate. The dependence of the Θ-temperature upon the state of the polyelectrolyte chains might conceivably be expressed by writing it as a function of any one of α, α' or pH. However, Flory and Osterheld[120] have argued that α is the most appropriate parameter since the Θ-point is then associated with a unique average chain composition, that is if the polyelectrolyte chain is regarded as a statistical copolymer of sodium acrylate and acrylic acid only α gives the copolymerisation ratio of the notional copolymer.

The effect of added salt and α on the flocculation of poly(styrene) latices stabilised by poly(acrylate) chains has been investigated by Buscall.[111] The effect of α is illustrated in Fig. 21, the medium in this

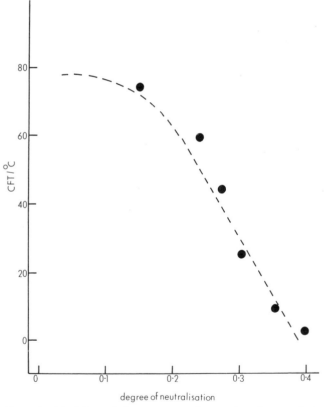

Fig. 21. Plot of critical flocculation concentration against degree of neutralisation, α, for latices stabilised by sodium polyacrylate chains. Background electrolyte concentration, $1 \cdot 6 \, \text{mol dm}^{-3}$; ●, CFT; – – –, Θ temperature.

instance being $1.6\,\text{mol dm}^{-3}$ sodium chloride. The effect on the Θ-point is also shown.[121] The correlation between CFT and Θ is very precise, a similar but somewhat less precise correlation was observed at lower and higher ionic strengths. Of particular interest was the effect of sodium iodide on the latices. $1:1$ electrolytes such as sodium chloride and sodium bromide lower the Θ-point of poly(acrylate) chains. The effect of sodium iodide is, however, more complex. The addition of sodium iodide to high molecular weight sodium poly(acrylate) causes liquid–liquid phase separation at a concentration of $\sim 1.3\,\text{mol dm}^{-3}$, remixing at $\sim 3\,\text{mol dm}^{-3}$ and a second phase separation at $\sim 6\,\text{mol dm}^{-3}$. The flocculation behaviour of poly(acrylate)-stabilised latices was found to parallel the solution behaviour inasmuch as latices were stable below $1.5\,\text{mol dm}^{-3}$ and between 2.6 and $6\,\text{mol dm}^{-3}$ but unstable at sodium iodide concentrations corresponding to the two-phase regions. This rather unusual flocculation behaviour clearly illustrates the connection between critical solution and incipient flocculation behaviour that exists for sterically stabilised systems.

It can be deduced from Fig. 21 that the concentration of sodium chloride required to cause flocculation at any particular temperature increases with α (except perhaps around $\alpha = 0.1$). The reverse dependence is observed with electrolytes containing ions that form insoluble salts with the polyelectrolyte since it is normally only the ionised chain that is capable of forming the complex. An example is the Ca^{2+}/poly(acrylate) system. The addition of calcium salts to solutions of sodium poly(acrylate) causes precipitation of the polymer at a calcium concentration which is somewhat lower but in proportion to the stoichiometric amount;[122] solutions of poly(acrylic acid), on the other hand, will withstand a substantial excess of Ca^{2+}. The flocculation of latices is found to be exactly analogous.[123] If, however, an insoluble complex is not formed, divalent ions may be no more effective in causing flocculation than $1:1$ electrolytes. Thus latices stabilised by sodium poly(α-methyl styrene sulphonate) have been found to remain stable in saturated calcium chloride.[83] From the limited amount of data available, it would appear that the flocculating power of simple cations for latices stabilised by anionic polyelectrolytes follows the order:

$$M_{ic}^{2+} \gg M^{2+} \approx M^+$$

where the subscript ic implies ions which undergo insoluble complex formation. The corresponding order for conventional charged latices is:

$$M_{ic}^{2+} \gtrsim M^{2+} \gg M^+$$

and for latices stabilised by non-ionic polymers typically:

$$M^+ \gtrsim M^{2+}$$

Polyelectrolyte-stabilised latices can thus show flocculation behaviour which is superficially similar to that shown by conventional charged latices if the added electrolyte complexes with the charged groups. Otherwise the behaviour is more akin to conventional sterically stabilised latices.

The experimental work discussed thus far in this section was performed using latices stabilised by adsorbed comb-graft polyelectrolytes. Polyelectrolyte-stabilised latices can also be prepared by dispersion polymerisation of a suitable vinyl monomer, for example styrene, in the presence of homopolymers provided a suitable grafting initiator is used.[117] The latices are stabilised by a graft copolymer produced *in situ*. Such latices tend to be more sensitive to added electrolytes than do latices stabilised by preformed graft copolymers, and can show very complicated flocculation behaviour.[117–119] Another method of producing latices with some polyelectrolyte character is to incorporate a water-soluble ionogenic monomer, for example acrylic acid, in a conventional emulsion polymerisation. The technological literature is full of examples of this, and well-characterised latices have been prepared by Bassett and Hoy.[124,125]

Only anionic polyelectrolytes have been considered here. Amphipathic cationic copolymers have been prepared and described as being good stabilisers and dispersants.[126,127] There do not, however, appear to have been any detailed studies of the incipient flocculation behaviour of latices stabilised by cationic polyelectrolytes. Amphoteric polyelectrolytes in the form of proteins have been studied both as stabilisers and flocculants, and the behaviour of these will be described in a subsequent section.[155]

3.1.9. Minimum Stabiliser Molecular Weight for Effective Stabilisation
In practical applications of steric stabilisation it is often desirable on both economic and synthetic grounds to keep the stabilising chains as short as possible. It is thus useful to attempt to establish at what lower level of molecular weight or degree of polymerisation effective stabilisation is lost. The available evidence will be reviewed in this section.

TABLE 8

Effect of Stabiliser Chain Length on the Flocculation Behaviour of Latices Stabilised by Poly(α-methyl styrene sulphonate) Chains[83]

$\bar{M}_w$	$\bar{M}_n$ (as α-methylstyrene)	$\overline{DP}$	CFC of sodium chloride (mol dm^{-3})
1 250	1 000	8	1·6
1 800	1 600	14	>5
2 900	2 000	17	>5
3 800	3 000	25	>5
9 100	7 000	60	>5
9 800	8 000	68	>5
14 000	12 000	102	>5

The effect of stabiliser chain length on the flocculation behaviour of polyelectrolyte-stabilised latices has been investigated by Corner and Gerrard[83] using a series of well-characterised comb-grafts. These consisted of a poly(methyl methacrylate) backbone onto which were grafted poly(α-methylstyrene sulphonate) side-chains of low polydispersity. The effectiveness of the copolymers as stabilisers was assessed by determining the CFC of sodium chloride for latices onto which the copolymers had been adsorbed. Some of the results are summarised in Table 8. At a degree of polymerisation ($\overline{DP}$) of the side-chains of 14 or above (MW $\gtrsim$ 3000) the latices were stable in the presence of saturated calcium and sodium chlorides. At a lower $\overline{DP}$ of 8, the latices could be flocculated by the addition of sodium chloride, but the CFC was still substantially higher than the ccc of the bare latex at 1·6 mol dm^{-3} as compared with ~0·2 mol dm^{-3}, showing that a measure of steric stabilisation still existed. The layer thickness was not determined, but as the contour length of vinyl chains is only ~0·3 nm per monomer unit, the layers must have been thin and much less than 5 nm at the lower degrees of polymerisation.

A similar study of a non-aqueous system has been made by Dawkins and Taylor[107] using poly(dimethyl siloxane) (PDMS) stabilising chains incorporated into AB block copolymers. The barrier thickness *was* measured and the results suggested that the PDMS chains adopted an extended conformation at $\overline{DP}$ < 150. PDMS is one of the most flexible chains known, and so it is tempting to speculate that short chains might, in general, adopt an extended conformation at interfaces; the structure of the copolymer and the surface concentration of chains may, however, have a very important effect. The range of $\overline{DP}$ covered by Dawkins and Taylor did not extend as low as in the work of Corner and Gerrard,[83] the lowest $\overline{DP}$ being ~43; the CFT were, however, found to be independent of $\overline{DP}$ over the range covered (cf. Table 5). PDMS was also used in the stabilising moiety by Everett and Stageman[128] who employed ABA block copolymers in which PDMS was the central block. In this instance, an effect of molecular weight on the CFT was observed for $\overline{DP}$ as high as 200 and so it would appear perhaps that much longer chains are required when the stabilising barrier consists of loops rather than tails. This could either be because for a given $\overline{DP}$ the barrier thickness has to be smaller in the case of loops or because the segment density in the periphery of the adsorbed layer is lower. Predominantly loopy adsorption is also to be expected when homopolymers and random copolymers are used as stabilisers, although now the size of the average loop is likely to be much smaller than the size of a single chain. It might thus be anticipated that a higher $\overline{DP}$ still would be required in order to impart full steric stabilisation. The effect of molecular weight on the CFT of poly(styrene) latices stabilised by poly(vinyl alcohol) (partially hydrolysed poly(vinyl acetate)) has been examined by Tadros and Winn.[129] Their results are shown in Fig. 22. The CFT initially increased with molecular weight (indicating an increase in stability; PVA has an LCST) until a value of 10^4 was reached ($\overline{DP} \sim 200$), thereafter it decreased somewhat and finally reached a constant, plateau value above 2×10^4 ($\overline{DP} \sim 400$). On the basis of the limited amount of information available, it would appear that the minimum molecular weight for effective stabilisation follows the order:

$$M \text{ (multiple loops)} > M \text{ (single loop)} > M \text{ (tail)}$$

Further work is, however, required in order to establish whether such generalisations can be made. The work of Corner and Gerrard[83] and others shows nevertheless that effective stabilisation can be obtained using chains of very low $\overline{DP}$ especially when well-defined comb-graft stabilisers are used.

3.1.10. The Effect of Particle Concentration on the CFT

It has been implied so far that provided the stabiliser is well anchored and its molecular weight is not too low flocculation should occur at a critical point that correlates approximately with the Θ-point for exactly similar chains in solution. There are, however, a number of specific

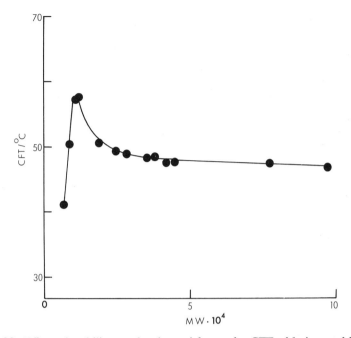

Fig. 22. Effect of stabiliser molecular weight on the CFT of latices stabilised by partially hydrolysed poly(vinyl alcohol).[129]

effects which can complicate this picture, including an occasionally observed dependence of the CFT on particle concentration.

The work of Long *et al.*[130] has shown that the intrinsically reversible nature of secondary-minimum flocculation and incipient flocculation should lead to a dependence of CFT (etc.) on particle concentration. Their argument embodies the idea that the free energy of floccule formation comprises two terms: an interaction term proportional to V_T (eqn (42)), and a configurational entropy term which accounts for the loss of particle translational entropy involved in forming the particles into a floc. It is evident that if internal motions in the floc are neglected the latter term is of order: $\mathbf{k} \ln (\phi/\phi_F)$ per particle, where ϕ is the mean volume fraction of solids in the dispersion and ϕ_F is that in a floc. The free energy of formation of a floc is thus approximately:

$$\Delta G_F \approx -\bar{N}\mathbf{k}T \ln (\phi/\phi_F) + \bar{N}\bar{C}V_T \qquad (60)$$

where $\bar{N}$ is the number of particles in a floc and $\bar{C}$ is the mean coordination number in the floc. Since $\phi < \phi_F$ for dilute systems, it is evident that the first term is positive and so it resists flocculation. Long et al.[130] have argued that provided the minimum in V_T is small, of the order of a few kT, then there is a critical volume fraction below which flocculation will not occur. It is evident from eqn (60) that this is of order:

$$\phi_{crit} \approx \phi_F \exp{(\bar{C}V_{min}/kT)} \sim \exp{(\bar{C}V_{min}/kT)} \sim 10^{-C |V_{min}|/2 \cdot 3kT}$$

(61)

There should thus be an effect of volume fraction on the CFT but at very small volume fractions, that is unless perhaps the particles are very small (since $V_{min} \sim a$, everything else being equal).

Incipient flocculation experiments are usually performed somewhere in the volume fraction range $10^{-4} < \phi < 10^{-2}$. Very often the CFT are found to be independent of ϕ; however, a pronounced effect is sometimes observed, a 10 K change in CFT in the direction of good solvency per decade increase in ϕ being not untypical. Such an effect has been observed both with small ($a < 0 \cdot 1 \mu$m) and large ($a > 1 \mu$m) particles,[119,128,129] and at volume fractions up to $0 \cdot 1$.[84] These observations make it unlikely that the effect is simply that described by Long et al.[130] At present then, there is no obvious reason why certain systems display a dependence of CFT on ϕ and others do not. However, in all cases reported, the molecular weight of the stabiliser was fairly low ($< 30\ 000$), and this may be significant;[84] the dispersion medium was also either a mixed or impure solvent, or an electrolyte solution, and the multicomponent nature of the medium may also be significant.[121] The existence of a dependence of CFT on latex concentration has led Vincent and Whittington[131] to consider the notion of a temperature–composition phase diagram for sterically stabilised latices somewhat analogous to the phase diagram for a polymer solution. Speculations as to the form of the complete phase diagram are to be found in their review.

3.1.11. Heterosteric Stabilisation

Two latices stabilised by chemically different stabilisers may either heteroflocculate or remain stable upon mixing. The theory of hetero-steric stabilisation and flocculation has been elaborated by Feigin and Napper.[157] The theory predicts that the particles will remain stable if the two stabilising polymers (1 and 2) are incompatible ($\chi_{12} > 0$) but heteroflocculate should the two polymers be compatible ($\chi_{12} < 0$), and particularly if the two polymers interact strongly (for example by H-bonding). Thus Napper[92] states that whilst it is possible to mix latices stabilised by poly(ethylene oxide) and poly(acrylamide) together without flocculation occurring, latices stabilised by poly(ethylene oxide) and poly(acrylic acid) heteroflocculate; poly(ethylene oxide) and poly(acrylamide) are incompatible whereas poly(ethylene oxide) and unionised polyacrylic acid form an H-bonded complex.

Feigin and Napper[157] also pointed out that it should be possible to flocculate one species in a stable mixture selectively, and that this is so has been demonstrated by Croucher and Hair.[105] Mixtures of particles stabilised by poly(styrene) (PS) and by poly(isobutylene) (PIB) in cyclopentane were found to be stable between the LCFT and UCFT observed for the PS-stabilised homodispersion, outside of this range the PS-stabilised particles flocculated leaving PIB-stabilised particles in dispersion. Polymer heterostabilisation and flocculation presumably have an important role to play in biological systems but these have yet to be investigated from this standpoint.

3.1.12. The Effect of Non-Ionic Surfactants on the Stability of Aqueous Latices

Whilst there is no clear distinction in kind between non-ionic surfactants which are used as stabilisers and polymeric surfactants or stabilisers the former term tends to be used to describe compounds of relatively low molecular weight such as alkanol and alkylphenol ethoxylates having oligomeric rather than polymeric oxyethylene head groups. It has been shown that such materials, of which dodecylhexaoxyethylene glycolmonoether $(C_{12}E_6)$ is an example, can adsorb from water onto, for example, poly(styrene) latices via their alkyl chains to give a hydrated monolayer of ethylene oxide chains that extend into the medium.[85]

If with a surfactant of the type $C_X E_Y$, Y is small, say $Y \ll 20$, then the steric barrier provided by adsorbed surfactant may well not be thick enough to prevent flocculation under the influence of van der Waals attraction. This appears to be so with $C_{12}E_6$ which has been studied by Ottewill and Walker.[85] The ccc of lanthanum nitrate for poly(styrene) latices was determined at various surface coverages of $C_{12}E_6$ and the results obtained for a latex with a particle diameter of 60 nm are illustrated in Fig. 23. The ccc increased with increasing coverage until at about 80% monolayer coverage flocculation could not be caused by the addition of $La(NO_3)_3$. However, the stabilising

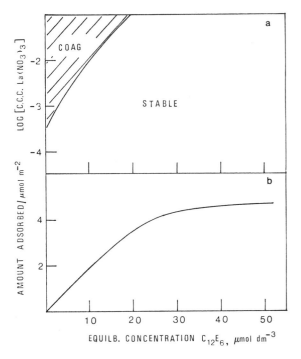

Fig. 23. The effect of adsorption density on the stability of poly(styrene) latices stabilised by $C_{12}E_6$.[85]

action was found to be less effective with larger particles (400 nm). This can be understood in terms of the difference in strength of the van der Waals attraction between the two latices. The energy of attraction between two spheres at close approach is given approximately by:

$$V_A \simeq -\frac{aA_{\text{eff}}}{12h} \tag{62}$$

For two particles separated by a steric barrier of thickness δ, this becomes:

$$V_A \simeq -\frac{aA_{\text{eff}}}{24\delta} \tag{63}$$

The composite Hamaker constant for poly(styrene)/water is close to $2kT$ and δ as obtained from sedimentation velocity measurements was

$5 \pm 0{\cdot}5$ nm at full coverage, and so for the smaller latex:

$$V_A \approx \frac{2{\cdot}30}{24{\cdot}5} = 0{\cdot}5\mathbf{k}T \tag{64}$$

and for the larger latex $V_A \approx 3{\cdot}3\mathbf{k}T$. There was thus a small but not insignificant van der Waals attraction at all coverages in the case of the larger latex.

Work on the effect of non-ionic surfactants on the stability of PTFE latices[156] has shown that the nature of the substrate can have an influence on stability. At high coverages the stabilising effect was similar to that obtained for polystyrene; however, at low coverages of $C_{12}E_6$ the ccc decreased rather than increased with coverage. It is known that hydrocarbon chains are reluctant to adsorb on PTFE surfaces[156] and so a probable explanation for this behaviour is that the initial adsorption occurs via the oxyethylene head group rather than the alkyl group giving pendant hydrocarbon chains which can then nucleate the formation of a second layer of surfactant.

The more important differences between stabilisation by low molecular weight non-ionic surfactants and high molecular weight polymers may be summarised as follows:

(1) Adsorption is reversible and so a standing concentration of surfactant has to be maintained in solution in order to ensure stability.

(2) There can be a stabilising effect well below full or saturated surface coverage whereas with polymeric stabilisers bridging flocculation (*et seq.*) will often occur.

(3) Stability is likely to depend upon a balance of V_A, V_R and V_S, not just upon V_S.

3.2. Latex Stability in the Presence of Free, Dissolved Polymer—Depletion Flocculation

The effect of adsorbed polymer layers on stability has so far been considered; however, theoretical calculations predict that free polymer in solution should also affect the stability of dispersed particles and cause weak flocculation under certain circumstances.[132-135] Any such effect would be difficult to demonstrate experimentally using bare latices and dissolved polymer since most real polymers have some propensity for adsorption. However, exactly the same effects are to be expected when soluble homopolymer is added to latices sterically stabilised by chemically identical chains that are irreversibly adsorbed or otherwise strongly anchored to the particle surface.[136-138]

Whether or not the addition of soluble polymer to sterically stabilised latices has any influence on their stability is found to depend upon the concentration and molecular weight of the added polymer. Typically weak flocculation is found to occur at some critical concentration of added polymer, followed by restabilisation at some higher concentration. Such behaviour can be seen in Fig. 24 which shows three-component domain diagrams for poly(ethylene oxide)-stabilised latices containing dissolved poly(ethylene oxide) of low (400) and high (10 000) molecular weights.[136] The location of the flocculation (shaded) region in respect of dissolved polymer concentration shows a pronounced inverse dependence upon molecular weight. In order to understand this behaviour it is necessary to look at the solution behaviour of polymers in a little detail.

It is well established that flexible polymers dissolved in good solvents exist as expanded coils which in terms of their interaction behave as nearly impenetrable spheres. This is so because overlap of two such coils is energetically highly unfavourable; thus interpenetration with no change in conformation would involve a partial demixing of polymer

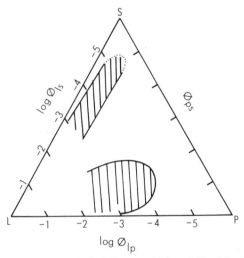

Fig. 24. Phase diagram for poly(ethylene oxide)-stabilised latices containing free, dissolved PEO. Upper diagram—MW = 10 000; lower diagram—MW = 400. L = latex; P = free PEO; S = aqueous medium.[136]

segments and solvent in the overlap zone, whereas a change in conformation in order to lower the segment density in the overlap region would involve a reduction of the configurational entropy of the chains. Polymer solutions are considered to be dilute in thermodynamic terms when the concentration is less than a critical value C_2^* where the coils start to pack.[139] C_2^* is of the order:

$$C_2^* \sim N/\langle r^2 \rangle^{\frac{3}{2}} \qquad (65)$$

or in terms of a volume fraction:

$$\phi_2^* \sim N^{-\frac{4}{5}} \qquad (66)$$

where N is the number of segments per chain and $\langle r^2 \rangle^{\frac{1}{2}}$ is their mean-square end-to-end length, or some other suitable measure of the coil dimensions. The second relationship follows from the first because $\langle r^2 \rangle^{\frac{1}{2}}$ is expected from theory to vary as $N^{\frac{3}{5}}$ in good solvents (strictly athermal solvents, i.e. $\chi = 0$). At concentrations above C_2^* polymer molecules have no choice but to overlap, and at some higher but less well-defined concentration C_2^{**} the segment density throughout the solution becomes quite uniform. For reasons similar to those outlined above, polymer molecules dissolved in the continuous phase of a sterically stabilised latex are unlikely to penetrate the adsorbed layer in the dilute regime. Further, if the adsorption is strong and/or the molecular weight of the adsorbed polymer is lower than that of the free polymer so that the segment density at the surface is high, significant penetration is also unlikely in the semi-dilute $(C_2 > C_2^*)$ regime. Under such circumstances the dispersed particles can be considered to behave as impenetrable hard spheres. Given this, it is evident that if two or more particles find themselves separated by a distance $H < 2R_G$, where R_G is the radius of gyration of the free polymer, the region between the particle has to be depleted of polymer. The osmotic pressure in this region will then be lower than in the bulk of the solution and the particles will experience a net force of attraction arising from the osmotic stress on the solvent. The semi-dilute regime $(C_2^* \leqslant C_2 \leqslant C_2^{**})$ has been analysed by Vincent et al.[136] using Flory–Huggins theory and by Gast et al.[158] using the mean-field theory of Muthukumar and Edwards.[159] They showed that the net free energy of mixing can change sign and make an attractive contribution to V_S at a critical concentration of dissolved polymer. The critical concentration is found to be dependent upon the polymer–solvent interaction parameter, the polymer molecular weight and the particle concentration; nevertheless the critical concentration is typically of similar order to C_2^* and, everything else being equal, shows a similar dependence on molecular weight. In this respect the theory appears to be in accord with the available experimental results.[137,138,140–142]

Gast et al.[158] used the liquid-state theory (second-order perturbation theory) to explore the effect of particle concentration on the state of dispersion. Their calculations were made for aqueous systems and so double layer repulsion was also included. Their results can be illustrated by a phase diagram of the type shown schematically in Fig. 25. Region I represents the stable region and Region II a co-existence region where flocs are in equilibrium with discrete particles. Region III represents a region of solid-like order at very high particle concentrations, the phase transition being caused by excluded-volume interaction between the particles. The restabilisation region, Region IV, requires further explanation. At high polymer concentrations, where overlap of the polymer coils has to occur, excluded volume (mixing) interaction between the polymers becomes screened as the segment density throughout the system becomes more and more uniform. Thus,

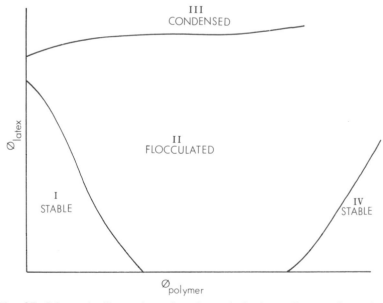

Fig. 25. Schematic illustration of a theoretical phase diagram for a latex containing free, dissolved polymer.

although the polymer mixing or osmotic contribution to the free energy of interaction of the particles remains net attractive at high polymer concentrations, its absolute magnitude decreases. This allows other sources of repulsion to take over and for the systems considered by Gast *et al.* the restabilisation is a consequence of double layer repulsion. Vincent and co-workers,[84,136] and Napper[92] have argued that the elastic contribution to V_S should also cause restabilisation since this is unaffected by screening, and in this sense stability in concentrated polymer solutions is akin to stability in polymer melts. This topic will now be considered.

3.3. Stability in Polymer Melts

Polymer-coated particles dispersed in a melt of a dissimilar polymer will normally be unstable as a result of the incompatibility of dissimilar polymers. Polymer-coated particles dispersed in a melt of chemically identical chains should, however, be stable; this result can be established in two equivalent ways. On the one hand a dispersion of particles in a polymer melt can be regarded as an extreme case of a dispersion in a concentrated solution. From this point of view stability arises because the mixing term is screened-out, leaving only the repulsive elastic term. Alternatively, a melt can be regarded as an athermal solvent of high molecular weight and thus high molar volume; the elastic and mixing terms in the equation are of similar order, except that:

$$\Delta G_{mix} \sim \frac{1}{V_{solvent}} \tag{67}$$

whereas

$$\Delta G_{elastic} \sim \frac{1}{V_{segment}} \tag{68}$$

In a melt $V_{solvent} \gg V_{segment}$, and so again it can be deduced that $|\Delta G_{mix}| \ll |G_{el}|$. The stability of particles in melts has been investigated by Smitham and Napper[143] who have shown that the behaviour of latices dispersed in liquid poly(ethylene oxide) is consistent with this picture.

3.4. Irreversible Flocculation by Soluble Polymer

3.4.1. Introduction

The addition of soluble polymers to latices can cause flocculation to occur in several ways. Two types of reversible flocculation, incipient flocculation and depletion flocculation, have already been discussed in the context of steric stabilisation. Reversible flocculation can also be caused by heating or cooling a latex containing dissolved polymer through the cloud point of the latter so that liquid–liquid phase separation of polymer solution forming the dispersion medium occurs. In such a case, flocculation involves the formation of liquid bridges between the particles, that is bridging of particles by lenses formed from the polymer-rich phase, although whether the bridges are formed from droplets which nucleate in solution or whether the particles act as nuclei for the spinodal decomposition is not clear. Soluble and dispersible polymers† can also cause irreversible flocculation. In order to cause irreversible flocculation, it is usually necessary for the polymer to adsorb on the particles (or vice versa if the polymer is large and the particles small); nevertheless, flocculation is often favoured by conditions of mixing and regimes of concentration that make likely encounters between particles that are only part-covered with adsorbed polymer. Thus many polymers capable of causing flocculation under certain appropriate circumstances, will act as steric stabilisers if full-coverage of the particles can be achieved. An obvious means of arranging collisions between partially covered particles is to add insufficient polymer to cover the particles, and doing so invariably causes flocculation. Factors such as the type and intensity of mixing, and the molecular weight of the soluble polymer can also have an important influence since they can affect the kinetics of adsorption, and so the competition between the particles for the polymer. If the molecular weight of the polymer is very high ($>10^6$), then it is not uncommon to find that flocculation occurs regardless of the polymer dosage and mixing conditions employed. This in part may be a consequence of the kinetics of adsorption, in that collisions between part-covered particles are always likely if the adsorption is sufficiently slow; however, it is important to realise that at high molecular weight the hydrodynamic radius of the polymer may well be larger than that of the particles. It may thus be more appropriate to think of binding of particles onto polymer molecules, rather than the reverse in such cases. Thus steric stabilisation may well be difficult to achieve on both kinetic and geometric grounds if the particles are small and the polymers large.

† Many commercial and experimental polymeric flocculants are ill-characterised materials of high molecular weight. Although these *appear* to be dissolvable, it is likely that many such materials exist in solution as swollen particles rather than as single molecules.

It will be evident from what has been said so far, that *bridging*[144] of two or more particles by adsorbed macromolecules is a plausible mechanism of irreversible polymer flocculation. There are, however, additional effects, and with polyelectrolytes and charged particles, charge neutralisation can also play a part.

3.4.2. Flocculation by Non-Ionic Polymers
An example of the flocculation of latices by non-ionic polymers is provided by the work of Ash and Clayfield[145] who examined the effect of adding poly(ethylene oxide) (PEO) on the flocculation of anionic poly(styrene) latices. The experiments were done in the presence of sufficient barium chloride to cause slow flocculation ($1 \ll W < 10^2$) in the absence of polymer. The addition of PEO caused the flocculation rate to go through a maximum where rapid flocculation ($W = 1$) was observed. The amount of PEO required to cause rapid flocculation decreased with increasing molecular weight, the range studied was 6×10^4 to 5×10^6 and the optimum PEO concentration halved over this range. This *sensitisation* to flocculation and the existence of an optimum dose of polymer are well known from work on the polymer flocculation of inorganic sols (see, for example, the reviews by Healy and La Mer[144] and Vincent[146]). The work of Ash and Clayfield is one of very few studies of flocculation by non-ionic polymers made using latices.

3.4.3. Flocculation of Charged Latices by Polyelectrolytes Having Like Charge
Polyelectrolytes can cause flocculation of latices having a similar electrostatic charge, although it is almost invariably found that divalent counterions need to be present for flocculation to occur. The precise mechanism has been the subject of some debate in the literature, although Vincent[146] in a detailed review of the subject concluded that 'Bridging is the predominant mechanism in flocculation by polyelectrolytes of the same sign as the particles', and that 'the general features are broadly similar to those occurring in flocculation by neutral polymers'. It has variously been suggested that the role of the divalent counterions is to form an interfacial complex between the particle surface and the charged polymer,[147] that the divalent ions simply screen the surface charge on the particles,[148] and that the species that actually bridges is a neutral divalent-ion/polyelectrolyte complex.[149] In view of the fact that divalent ions commonly form

strong complexes with soluble polyelectrolytes, the latter is a plausible suggestion. Strong support for the idea that flocculation involves a neutral or near-neutral complex comes from the work of Lindström and Söremark[150] on the flocculation of negative polymer particles by sodium alginate. Only those cations that bound strongly to sodium alginate were found to be effective as coflocculants.

3.4.4. Flocculation of Charged Latices by Polyelectrolytes of Dissimilar Charge

The general pattern of flocculation at low polymer doses and either restabilisation or less efficient flocculation at higher doses is also observed where the polymer and particle have unlike charge. Charge neutralisation is, however, undoubtedly involved in the mechanism of flocculation.

The addition of cationic polymers to anionic latices has been studied by Kasper,[151] Gregory[152] and Ishikowa.[153] It has been noted by Gregory that the optimum dose of polyelectrolyte is often similar to that required to reduce the particle electrophoretic mobility to zero. Flocculation thus appears to occur close to the isoelectric point of the particle–polymer complex. It is also found that the optimum dose is independent of the molecular weight. These observations clearly suggest that charge neutralisation is important to, even if it is not the actual mechanism of, flocculation. Kasper and Gregory have emphasised that neutralisation of the charge overall does not necessarily imply that each and every surface anionic group is neutralised by a cationic group. Rather, since this would require the flocculating polymer to unfold and map each charged site on the surface, it is much more likely that the charge is only neutralised in a statistical sense. The neutral surface is thus imagined as consisting of a mosaic of positive (over-neutralised) and negative (under-neutralised) regions on a scale dictated by the size of the cationic polymer molecules. There should be an orientation-dependent force of attraction between particles that are charged-neutralised in this sense[151] and so some support for the 'charge-patch' model can be gained from the observation[152] that the rate of flocculation at optimum dosing is faster than the Smoluchowski fast rate (implying a source of long-range attraction). Further, the same flocculation rate was obtained when the overdosed (+ve) and under-dosed (−ve) particles were mixed. These observations are hard to explain without invoking charge heterogeneity. It has also been found that many of the features of flocculation by cationic polyelectrolytes can be observed if the soluble polyelectrolyte is replaced by small cationic latices having a particle size much smaller than that of the anionic latex.[154] This similarity between polymer flocculation and heteroflocculation implies that in the former case extensive unfolding does not occur. In the soluble polyelectrolyte case, restabilisation is observed at high polymer doses, but the breadth of flocculation region is found to increase with molecular weight. By contrast, the flocculation region observed in the case of flocculation by cationic surfactants[42] was very sharp; in this instance there is little possibility of a charge-patch interaction, and so the existence of a broad flocculation region in the polymeric case might also be taken as evidence in support of the charge-patch model.

3.4.5. The Effect of Amphoteric Polyelectrolytes (Globular Proteins) on Stability

Work on proteins provides a rather special example of flocculation by amphoteric polyelectrolytes. The effect of two soluble proteins, human serum albumin (HSA) and human fibrinogen (HF6), on the stability of poly(styrene) latices has been investigated by van der Scheer *et al.*[155] The two proteins were found to act as steric stabilisers at high doses and as polymeric flocculants at lower doses. The proteins had accessible isoelectric points at pH 4·9 (HSA) and pH 5·5 (HF6), and so it was possible to vary both the sign and magnitude of the overall charge carried by the proteins. The effects of electrolyte, protein concentration and pH on the kinetics of flocculation were examined, and it was possible to distinguish many of the various effects stabilising and destabilising described in the preceding sections in one system. It is probably possible to do no better here than to reiterate some of the conclusions of van der Scheer *et al.* They concluded:

(1) 'The adsorbed molecules protect (sterically-stabilise) the particles against aggregation by electrolyte addition if they occupy the surface of the latex particles completely, and if the continuous phase has 'good solvent' properties for the protein.'

(2) 'No steric stabilisation is observed when the continuous phase has 'bad solvent' properties for the protein.'

(3) 'Positively-charged protein molecules can induce flocculation of the negatively-charged latex by charge-neutralisation. At relatively low protein concentrations, the charge on the latex particles is reversed by adsorption of positive protein molecules, resulting in restabilisation of the latex by electrostatic repulsion.'

(4) 'The latex cannot be flocculated by negatively-charged protein molecules in the absence of salt. At low surface coverage, flocculation can be caused by very small additions of electrolyte (sensitisation by a bridging mechanism).'

These observations show that, as stabilisers and flocculants, amphoteric proteins combine many of the properties of simple synthetic polymers.

REFERENCES

1. Verwey, E. J. W. and Overbeek, J. Th. G., *Theory of stability of lyophobic colloids*, Elsevier, Amsterdam (1948).
2. Stern, O., *Z. Elektrochem.*, **30,** 568 (1924).
3. Kruyt, H. R., *Colloid science*, Vol. I, Elsevier, Amsterdam (1952).
4. Hunter, R. J., *Zeta potential in colloid science*, Academic Press, London (1981).
5. Wiersema, P. H., Loeb, A. and Overbeek, J. Th. G., *J. Colloid and Interface Sci.*, **22,** 78 (1966).
6. Ottewill, R. H. and Shaw, J. N., *J. Electroanal. Chem. and Interfacial Electrochem.*, **37,** 133 (1972).
7. O'Brien, R. W. and White, L. R., *J. Chem. Soc. Faraday Trans. II*, **74,** 1607 (1978).
8. Hearn, J., Ottewill, R. H. and Shaw, J. N., *Brit. Polymer J.*, **2,** 116 (1970).
9. Reerink, H. and Overbeek, J. Th. G., *Disc. Faraday Soc.*, **18,** 74 (1954).
10. Hamaker, H. C., *Physica*, **4,** 1058 (1937).
11. Kitchener, J. A. and Schenkel, J. H., *Trans. Faraday Soc.*, **56,** 161 (1960).
12. Dzyaloshinskii, I. E., Lifshitz, E. M. and Pitaevskii, L. P., *Advances in Physics*, **10,** 165 (1961).
13. Mahanty, J. and Ninham, B. W., *Dispersion forces*, Academic Press, London (1976).
14. Richmond, P., in *Colloid science*, Vol. 2, D. H. Everett (ed.), Royal Society of Chemistry, London (1975).
15. Evans, R. and Napper, D. H., *J. Colloid and Interface Sci.*, **45,** 138 (1973).
16. Hough, D. B. and White, L. R., *Advances in Colloid and Interface Science*, **14,** 3 (1980).
17. Derjaguin, B. V. and Landau, L., *Acta Physicochim. USSR*, **14,** 633 (1941).
18. Fowkes, F. M., *Ind. Eng. Chem.*, **56,** 40 (1964).
19. Schulze, H., *J. Prakt. Chem.*, **25,** 431 (1882); **27,** 320 (1883).
20. Hardy, W. B., *Proc. Roy. Soc.*, **A66,** 110 (1900).
21. von Smoluchowski, M., *Z. Phys. Chem.*, **92,** 129 (1917).
22. Fuchs, N., *Z. Phys.*, **89,** 736 (1934).

23. Overbeek, J. Th. G., *Colloid science*, Vol. 1, H. R. Kruyt (ed.), Elsevier, Amsterdam (1952).
24. Ottewill, R. H. and Shaw, J. N., *Disc. Faraday Soc.*, **42**, 154 (1966).
25. Derjaguin, B. V. and Muller, V. M., *Doklady Phys. Chem.*, **176**, 738 (1967).
26. Honig, E. P., Roeberson, G. J. and Wiersema, P. H., *J. Colloid and Interface Sci.*, **36**, 97 (1971).
27. Ottewill, R. H. and Rance, D. G., *Croatica Chem. Acta*, **50**, 65 (1977).
28. Ottewill, R. H. and Walker, T., *Kolloid Z. u. Z. Polymere*, **227**, 108 (1968).
29. Storer, C. C., PhD Thesis, University of Bristol (1968).
30. Pelton, R., PhD Thesis, University of Bristol (1976).
31. Neimann, R. E. and Lyashenko, O. A., *Colloid J. USSR*, (English trans.), **24**, 433 (1962).
32. Force, C. G. and Matijević, E., *Kolloid Z. u. Z. Polymere*, **224**, 51 (1968).
33. Bibeau, A. A. and Matijević, E., *J. Colloid and Interface Sci.*, **43**, 330 (1973).
34. Force, C. G. and Matijević, E., *Kolloid Z. u. Z. Polymere*, **225**, 33 (1968).
35. Matijević, E., *J. Colloid and Interface Sci.*, **58**, 374 (1977).
36. Kratohvil, S. and Matijević, E., *J. Colloid and Interface Sci.*, **57**, 104 (1976).
37. Cornell, R. M., Goodwin, J. W. and Ottewill, R. H., *J. Colloid and Interface Sci.*, **71**, 254 (1979).
38. Kayes, J. B., *J. Colloid and Interface Sci.*, **56**, 426 (1976).
39. Cebula, D. J., Thomas, R. K., Harris, N. M., Tabony, J. and White, J. W., *Faraday Disc. Chem. Soc.*, **65**, 76 (1978).
40. Harris, N. M., Ottewill, R. H. and White, J. W., *Adsorption from solution*, Academic Press, London (1983).
41. Bee, H. E., Ottewill, R. H., Rance, D. G. and Richardson, R. A., *Adsorption from solution*, Academic Press, London (1983).
42. Connor, P., PhD Thesis, University of Bristol (1968).
43. Ottewill, R. H., *Emulsion polymerization*, I. Piirma (ed.), Academic Press, New York (1982).
44. Connor, P. and Ottewill, R. H., *J. Colloid and Interface Sci.*, **37**, 642 (1971).
45. Ottewill, R. H., Rastogi, M. C. and Watanabe, A., *Trans. Faraday Soc.*, **56**, 854 (1960).
46. Rendall, H. M., Smith, A. L. and Williams, L. A., *J. Chem. Soc. Faraday Trans. I*, **75**, 669 (1979).
47. Storer, C. S., PhD Thesis, University of Bristol (1968).
48. Levine, S. and Bell, G. M., *J. Colloid Sci.*, **20**, 695 (1965).
49. Hogg, R., Healy, T. W. and Fuerstenau, D. W., *Trans. Faraday Soc.*, **62**, 1638 (1966).
50. Cheung, W. K., PhD Thesis, University of Bristol (1979).
51. Marshall, J. K. and Kitchener, J. A., *J. Colloid and Interface Sci.*, **22**, 342 (1966).
52. Hull, M. and Kitchener, J. A., *Trans. Faraday Soc.*, **65**, 3093 (1969).

53. Levich, V. G., *Physico-chemical hydrodynamics*, Prentice Hall, Englewood Cliffs, New Jersey (1962).
54. Clint, G. E., Clint, J. H., Corkill, J. M. and Walker, T., *J. Colloid and Interface Sci.*, **44**, 121 (1973).
55. Heller, W. and Peters, J., *J. Colloid and Interface Sci.*, **32**, 592 (1970); **33**, 578 (1970).
56. Heller, W. and de Lauder, W. B., *J. Colloid and Interface Sci.*, **35**, 60 (1971).
57. Heller, W. and de Lauder, W. B., *J. Colloid and Interface Sci.*, **35**, 308 (1971).
58. Frens, G. and Overbeek, J. Th. G., *J. Colloid and Interface Sci.*, **36**, 286 (1971).
59. Frens, G., PhD Thesis, University of Utrecht (1968).
60. Smitham, J. B., Gibson, D. V. and Napper, D. H., *J. Colloid and Interface Sci.*, **45**, 211 (1973).
61. Dunn, A. S. and Chong, L. C. H., *Brit. Polymer J.*, **2**, 49 (1970).
62. Fitch, R. M., *Polymer colloids II*, Plenum Press, New York (1980).
63. Goodwin, J. W., Ottewill, R. H., Pelton, R., Vianello, G. and Yates, D. E., *Brit. Polymer J.*, **10**, 173 (1978).
64. Goodwin, J. W., Hearn, J., Ho, C. C. and Ottewill, R. H., *Colloid and Polymer Sci.*, **60**, 173 (1976).
65. de Vries, A. J., *Rheology of emulsions*, P. Sherman (ed.), Pergamon Press, London (1963).
66. Zeichner, G. R. and Schowalter, W. R., *A. I. Chem. E. J.*, **23**, 243 (1977).
67. Zeichner, G. R. and Schowalter, W. R., *J. Colloid and Interface Sci.*, **71**, 237 (1979).
68. van der Ven, T. G. M. and Mason, S. G., *J. Colloid and Interface Sci.*, **57**, 535 (1976).
69. van der Ven, T. G. M. and Mason, S. G., *Colloid and Polymer Sci.*, **255**, 468 (1977).
70. Adler, P., *J. Colloid and Interface Sci.*, **83**, 106 (1981).
71. Hiltner, P. A. and Krieger, I. M., *J. Physical Chem.*, **73**, 2386 (1969).
72. Hachisu, S., Kobayashi, Y. and Kose, A., *J. Colloid and Interface Sci.*, **42**, 342 (1973).
73. Goodwin, J. W., Ottewill, R. H. and Parentich, A., *J. Physical Chem.*, **84**, 1580 (1980).
74. Ottewill, R. H., *Colloidal dispersions*, J. W. Goodwin (ed.), Royal Society of Chemistry, London (1982).
75. Cebula, D. J., Goodwin, J. W., Jeffrey, G. C., Ottewill, R. H., Parentich, A. and Richardson, R. A., *Faraday Disc. Chem. Soc.*, **76**, 37 (1983).
76. Hayter, J. B., Pynn, R. and Suck, J. B., *J. Phys. F.: Met. Phys.*, **13**, L1 (1983).
77. Alexander, K. A., Cebula, D. J., Goodwin, J. W., Ottewill, R. H. and Parentich, A., *Colloids and Surfaces*, **7**, 233 (1983).
78. Cebula, D. J., Goodwin, J. W., Ottewill, R. H., Jenkin, G. and Tabony, J., *Colloid and Polymer Sci.*, **261**, 555 (1983).
79. Tadros, Th. F., *Advances in Colloid and Interface Sci.*, **12**, 141 (1980).

80. Bensley, C. N. and Hunter, R. J., *J. Colloid and Interface Sci.*, **88,** 546 (1982).
81. Barrett, K. E. J. (ed.), *Dispersion polymerisation in organic media*, Wiley, New York (1975).
82. Schick, M. J. (ed.), *Non-ionic surfactants*, Edward Arnold Publishers, London (1967).
83. Corner, T. and Gerrard, J., *Colloids and Surfaces*, **5,** 187 (1982).
84. Vincent, B. and Whittington, S. G., in *Surface and colloid science*, Vol. 12, E. Matijević (ed.), Plenum, New York (1982).
85. Ottewill, R. H. and Walker, T., *Kolloid Z. u. Z. Polymere*, **227,** 108 (1968).
86. Napper, D. H., *J. Colloid and Interface Sci.*, **58,** 390 (1977).
87. Hesselink, F. Th., Vrij, A. and Overbeek, J. Th. G., *J. Phys. Chem.*, **75,** 2094 (1971).
88. Flory, P., *Principles of polymer chemistry*, Cornell University Press, Ithaca (1971).
89. Dolan, A. K. and Edwards, S. F., *Proc. R. Soc. Lond.*, **A343,** 427 (1975).
90. Gerber, P. R. and Moore, M. A., *Macromolecules*, **10,** 476 (1977).
91. Levine, S., Thomlinson, B. B. and Robinson, K., *Faraday Disc. Chem. Soc.*, **65,** 202 (1978).
92. Napper, D. H., in *Colloidal dispersions*, J. W. Goodwin (ed.), The Royal Society of Chemistry, London (1982), pp. 99–128.
93. Smitham, J. B., Evans, R. and Napper, D. H., *J. Chem. Soc. Faraday Trans. I*, **71,** 285 (1975).
94. Smitham, J. B. and Napper, D. H., *Colloid and Polymer Sci.*, **257,** 748 (1979).
95. Hesselink, F. Th., *J. Poly. Sci. Poly. Symp.*, **61,** 439 (1977).
96. Bagchi, P., *J. Colloid and Interface Sci.*, **50,** 115 (1975).
97. Di Margio, E. A. and McCrackin, F. L., *J. Chem. Phys.*, **43,** 539 (1965).
98. Croucher, M. D. and Hair, M. L., *Macromolecules*, **11,** 874 (1978).
99. Doroszkowski, A. and Lambourne, R., *J. Poly. Sci. part C*, **34,** 253 (1971).
100. Cairns, R. J. R., Ottewill, R. H., Osmond, D. W. J. and Wagstaff, I., *J. Colloid and Interface Sci.*, **54,** 45 (1976).
101. Lyklema, J. and van Vliet, T., *Faraday Disc. Chem. Soc.*, **65,** 25 (1978).
102. Cain, F. W., Ottewill, R. H. and Smitham, J. B., *Faraday Disc. Chem. Soc.*, **65,** 33 (1978).
103. Klein, J., *Advances in Colloid and Interface Sci.*, **16,** 101 (1982).
104. Croucher, M. D. and Hair, M. L., *J. Colloid and Interface Sci.*, **81,** 257 (1981).
105. Croucher, M. D. and Hair, M. L., *Colloids and Surfaces*, **1,** 349 (1980).
106. Croucher, M. D. and Hair, M. L., *Macromolecules*, **11,** 874 (1978).
107. Dawkins, J. V. and Taylor, G., *Colloid Polymer Sci.*, **258,** 79 (1980).
108. Cowell, C., Li-In-On, F. K. R. and Vincent, B., *J. Chem. Soc. Faraday Trans. I*, **74,** 337 (1978).
109. Napper, D. H., *J. Colloid and Interface Sci.*, **32,** 106 (1970).
110. Evans, R., Davison, J. B. and Napper, D. H., *Polymer Lett.*, **10,** 449 (1972).

111. Buscall, R., *J. Chem. Soc. Faraday Trans. I*, **77**, 909 (1981).
112. Evans, R., Data taken from Ref. 92.
113. Cornet, C. F. and van Ballegooijen, *Polymer*, **7**, 293 (1966).
114. Napper, D. H., *J. Colloid and Interface Sci.*, **33**, 384 (1970).
115. Ataman, M. and Boucher, E. A., *J. Poly. Sci.*, **20**, 1585 (1982).
116. Heller, W. and Pugh, T. L., *J. Chem. Phys.*, **24**, 1107 (1956); *J. Poly. Sci.*, **47**, 219 (1960).
117. Corner, T., *Colloids and Surfaces*, **3**, 119 (1981).
118. Buscall, R. and Corner, T., *Emulsion polymers and emulsion polymerisation*, D. R. Bassett and A. E. Hamielec (eds), ACS Symp. Ser. 165 (1981), p. 157.
119. Buscall, R. and Corner, T., *Colloids and Surfaces*, **5**, 333 (1982).
120. Flory, P. J. and Osterheld, J. E., *J. Phys. Chem.*, **58**, 653 (1954).
121. Buscall, R. and Corner, T., *Eur. Poly. J.*, **18**, 967 (1982).
122. Inegami, A. and Imai, N., *J. Polymer Sci.*, **56**, 133 (1962).
123. Corner, T. and Weise, G. R., Unpublished results; Buscall, R., Unpublished observations.
124. Hoy, K. L., *J. Coat. Tech.*, **51**, 27 (1979).
125. Bassett, D. R. and Hoy, K. L., in *Polymer Colloids II*, R. M. Fitch (ed.), Plenum Press, New York (1980), pp. 1–25.
126. Wessling, R. A. and Pickelman, D. A., *J. Dispersion Sci. and Technol.*, **2**, 281 (1981).
127. Marie, P., Hervenschmidt, Y-Le. and Gallot, Y., *Makromol. Chem.*, **177**, 2773 (1976).
128. Everett, D. H. and Stageman, J. F., *Faraday Disc. Chem. Soc.*, **65**, 231 (1978).
129. Tadros, Th. F. and Winn, P., in *The effect of polymers on dispersion properties*, Th. F. Tadros (ed.), Academic Press, London (1982).
130. Long, J. A., Osmond, D. W. J. and Vincent, B., *J. Colloid and Interface Sci.*, **42**, 545 (1973).
131. Vincent, B. and Whittington, S. G., in *Surface and colloid science*, Vol. 12, E. Matijević (ed.), Plenum Press, New York (1982).
132. Asakura, S. and Oosawa, F., *J. Poly. Sci.*, **33**, 183 (1958).
133. Vrij, A., *Pure Appl. Chem.*, **48**, 471 (1976).
134. Joanny, J. F., Liebler, L. and de Gennes, P. G., *J. Poly. Sci.*, **17**, 1073 (1979).
135. Feigin, R. I. and Napper, D. H., *J. Colloid and Interface Sci.*, **74**, 567 (1980); **75**, 525 (1980).
136. Vincent, B., Luckham, P. F. and Waite, F. A., *J. Colloid and Interface Sci.*, **73**, 508 (1980).
137. Li-In-On, F. K. R., Vincent, B. and Waite, F. A., ACS Symp. Ser. 9 (1975), p. 165.
138. Cowell, C., Li-In-On, F. K. R. and Vincent, B., *J. Chem. Soc. Faraday Trans. I*, **74**, 337 (1978).
139. de Gennes, P. G., *Scaling concepts in polymer physics*, Cornell University Press, New York (1979).
140. de Hek, H. and Vrij, A., *J. Colloid and Interface Sci.*, **70**, 592 (1979).
141. Clarke, J. and Vincent, B., *J. Colloid and Interface Sci.*, **82**, 208 (1981).

142. Sperry, P. R., Hopfenberg, H. R. and Thomas, N. L., *J. Colloid and Interface Sci.*, **82,** 62 (1981).
143. Smitham, J. B. and Napper, D. H., *J. Colloid and Interface Sci.*, **54,** 467 (1976).
144. Healy, T. W. and La Mer, V. K., *Rev. Pure Appl. Chem.*, **13,** 112 (1963).
145. Ash, S. G. and Clayfield, E. J., *J. Colloid and Interface Sci.*, **55,** 645 (1976).
146. Vincent, B., *Adv. Colloid Interface Sci.*, **4,** 193 (1974).
147. Sommerauer, A., Sussmann, D. L. and Stumm, W., *Kolloid Z.*, **225,** 147 (1968).
148. Kuz'kin, S. K. and Nebera, V. P., Cited in Ref. 146.
149. Sarkar, N. and Teot, A. S., *J. Colloid and Interface Sci.*, **43,** 370 (1973).
150. Lindström, T. and Söremark, C., *J. Colloid and Interface Sci.*, **60,** 258 (1977).
151. Kasper, D. R., PhD Thesis, CalTech (1971).
152. Gregory, J., *J. Colloid and Interface Sci.*, **42,** 448 (1973); **55,** 35 (1976).
153. Ishikowa, M., *J. Colloid and Interface Sci.*, **56,** 604 (1976).
154. Vincent, B., Young, C. A. and Tadros, Th. F., *Faraday Disc. Chem. Soc.*, **65,** 296 (1978).
155. van der Sheer, A., Tanke, M. A. and Smolders, C. A., *Faraday Disc. Chem. Soc.*, **65,** 264 (1978).
156. Bee, H. E., Ottewill, R. H., Rance, D. G. and Richardson, R. A., in *Adsorption from solution,* R. H. Ottewill, C. H. Rochester and A. L. Smith (eds), Academic Press, London (1983).
157. Feigin, R. I. and Napper, D. H., *J. Colloid and Interface Sci.*, **67,** 127 (1978).
158. Gast, A. P., Hall, C. K. and Russel, W. B., *Faraday Disc. Chem. Soc.*, **76,** 189 (1983).
159. Muthukumar, M. and Edwards, S. F., *J. Chem. Phys.*, **76,** 2720 (1982).
160. Buscall, R., Goodwin, J. W., Hawkins, M. W. and Ottewill, R. H., *J. Chem. Soc. Faraday Trans. I*, **78,** 2873 (1982).
161. Buscall, R., Goodwin, J. W., Hawkins, M. W. and Ottewill, R. H., *J. Chem. Soc. Faraday Trans. I*, **78,** 2889 (1982).

Chapter 6

Rheology of Polymer Colloids

I. M. KRIEGER

Case Western Reserve University, Cleveland, Ohio, USA

1. SOME DEFINITIONS

Rheology has been defined as the study of deformation and flow of matter. Colloidal dispersions, in which the particles are large compared to molecules of the dispersing liquid and yet small enough to exhibit Brownian movement, are particularly interesting to the rheologist. The simplest and most tractable of dispersions are the polymer colloids, since their particles are spherical and their interparticle forces controllable.

Although other modes of deformation and flow can be important, polymer colloids are ordinarily studied in shear flow. The important rheological variables can be defined by reference to the idealised instrument of Fig. 1, in which the sample under study is confined between parallel planes of area A. The planes are normal to the z-axis and are separated by a distance h which is small compared to the linear dimensions of the planes. When a force F directed parallel to the x-axis is applied to the upper plane, while the lower plane is held stationary, the test material experiences a uniform displacement gradient, as shown in Fig. 1. The *shear stress* σ is the force per unit area:

$$\sigma = F/A \tag{1}$$

while the *shear strain* γ is the displacement gradient:

$$\gamma = \frac{\mathrm{d}x}{\mathrm{d}z} \tag{2}$$

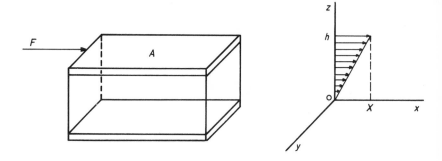

Fig. 1. Idealised parallel-plane rheometer.

If the material is an elastic solid, a constant strain γ_f will prevail until the stress is removed, after which the material will return to its initial state. The *shear modulus* G is the stress–strain ratio:

$$G = \frac{\sigma}{\gamma} \tag{3}$$

Also significant is the *stored energy density* w:

$$w = \int \sigma \, d\gamma \tag{4}$$

The idealised elastic solid obeys Hooke's law in shear, whereby the shear stress is proportional to the shear strain. A *Hookean solid* is therefore one whose shear modulus G is a material constant, depending only upon thermodynamic variables (temperature, pressure and composition). For the Hookean solid:

$$w = \tfrac{1}{2}G\gamma^2 \tag{5}$$

When the material confined between the planes is a fluid, the deformation increases with time. The shear stress determines the magnitude of the *shear rate* $\dot{\gamma}$:

$$\dot{\gamma} = \frac{du}{dz} = \frac{U}{h} \tag{6}$$

where u is the fluid velocity and $U = X$ is the velocity of the upper plane. The ratio of shear stress to shear rate is the *viscosity* η:

$$\eta = \frac{\sigma}{\dot{\gamma}} \tag{7}$$

The energy supplied by the force F is dissipated as heat within the fluid. The *power dissipation density* $\dot{w}$ is:

$$\dot{w} = \sigma\dot{\gamma} = \eta\dot{\gamma}^2 \tag{8}$$

Newton's law of viscous flow postulates proportionality between shear stress and shear rate. A *Newtonian fluid* is therefore one whose viscosity is a material constant, depending on temperature, pressure and composition.

In studying the rheology of polymer colloids, the appropriate concentration variable for the particles is their *volume fraction* ϕ, which is the dimensionless ratio of the volume of the particles to that of the entire dispersion. An important dimensionless viscosity variable is the *relative viscosity* η_r, which is the ratio of the viscosity η of the dispersion to that of the dispersing liquid, η_0:

$$\eta_r = \frac{\eta}{\eta_0} \tag{9}$$

2. INTRINSIC VISCOSITY

The intrinsic viscosity $[\eta]$ represents the contribution of an isolated particle to the viscosity of the dispersion. Its value depends mainly on the particle's shape and deformability. The intrinsic viscosity is determined experimentally as the initial slope of the η_r versus ϕ curve or, more sensitively, by extrapolating to $\phi = 0$ the ratio of $\ln \eta_r$ to ϕ. To calculate $[\eta]$ theoretically, it is necessary to solve the Navier–Stokes equations of fluid mechanics for slow flows, with boundary conditions appropriate to steady shear flow of an infinite medium containing an isolated particle. The result is a map of fluid velocities in the neighbourhood of the particle. This can be used to obtain the stresses on the particle surface, or to integrate the power dissipation density. Either route leads to the intrinsic viscosity.

The first calculation of an intrinsic viscosity was carried out by Einstein,[1] as part of his doctoral dissertation on Brownian motion. He found that the intrinsic viscosity for rigid spheres in a Newtonian medium is precisely 5/2, independent of their size. Einstein's result, obtained by calculating surface stresses, was later reproduced by Frisch and Simha[2] from integration of the power dissipation density. There have been numerous experimental verifications of Einstein's prediction. Intrinsic viscosities have been calculated theoretically for various particles other than rigid spheres. Ellipsoids of revolution are characterised by an axial ratio $r = b/a$. Here a is the length of the axis of revolution, while b is the length of the other axis of symmetry. Thus an oblate ellipsoid is one for which $0 < r < 1$, while a prolate ellipsoid has $r > 1$. A very oblate ellipsoid $(r \ll 1)$ resembles a disc; a very prolate ellipsoid $(r \gg 1)$ resembles a rod. Both Simha[3] and Kuhn and Kuhn[4] published expressions relating $[\eta]$ to r. For $r < 1$, Kuhn and Kuhn obtained:

$$[\eta] = \frac{5}{2} + \frac{32}{15\pi} \left(\frac{1}{r} - 1 \right) - 0 \cdot 628 \left(\frac{1-r}{1-0 \cdot 075r} \right) \tag{10}$$

and for $1 < r < 15$ they obtained:

$$[\eta] = \tfrac{5}{2} + 0 \cdot 4075(r-1)^{1 \cdot 508} \tag{11}$$

Simha's asymptotic expression for very large r (>15) applies both to ellipsoids and cylindrical rods:

$$[\eta] = \frac{14}{15} + \frac{r^2}{5} \left[\frac{1}{3(\ln(2r) - \lambda)} + \frac{1}{m(2r) - \lambda + 1} \right] \tag{12}$$

Here the parameter takes the value $1 \cdot 5$, while for rods $\lambda = 1 \cdot 8$. As r approaches infinity, $[\eta]$ approaches $0 \cdot 52r^2$.

Polymer latices are usually highly spherical in shape, and hence one would expect intrinsic viscosities of precisely 5/2, following Einstein's prediction. There are two complicating effects, however. One is the frequent presence of an adsorbed layer of surfactant, which increases the effective volume fraction of the colloid by a factor $(1 + d/a)^3$, where a is the particle radius and d the thickness of the adsorbed layer. The other is the primary electroviscous effect, wherein motion of the electrical double layer around a charged particle augments the energy dissipation during viscous flow.

Several studies which carefully included the effect of the adsorbed surfactant layer have reported intrinsic viscosities in the range 2·5–2·8.[5,6] When the particles are known to be perfectly spherical and the adsorbed layer irreversibly bound to the particle, the measured intrinsic viscosity can be used to calculate the effective thickness d of the adsorbed layer:

$$d/a = (2[\eta]/5 - 1)^{\frac{1}{3}} \qquad (13)$$

The particles in aqueous polymer colloids are electrically charged, and are surrounded by an 'ionic atmosphere' in which ions of opposite charge prevail. Even at high dilution, the interaction of the ion and its atmosphere augments the energy dissipation, and thereby increases the intrinsic viscosity. This 'primary electroviscous effect' was calculated first by von Smoluchowski,[7] and more recently by Booth.[8] Letting ζ be the zeta potential, ε the electronic charge and $\mathbf{k}$ Boltzmann's constant, the intrinsic viscosity is:

$$[\eta] = \tfrac{5}{2}\left[1 + \sum_{n=0}^{\infty} b_n \left(\frac{\varepsilon\zeta}{\mathbf{k}T}\right)^n\right] \qquad (14)$$

The coefficients b_n depend on the ionic strength and dielectric constant of the medium. The primary electroviscous effect produces only a small increment in viscosity, and is therefore difficult to measure. This may be the reason why recent studies by McDonogh and Hunter[9] show results which are not fully consistent with theory.

Distortions of shape could also cause $[\eta]$ to exceed the Einstein value. Doublets or higher particle aggregates could be present. Polymer colloids prepared by overpolymerising a second monomer onto the first are often found by electron microscopy to be non-spherical. Whatever the cause, the measured intrinsic viscosity would exceed $5/2$, since this represents a minimum for a rigid particle. Elastomeric polymer colloids will, in principle, deform under shear flow, leading to a dependency of intrinsic viscosity on shear stress. However, the stresses realised at attainable shear rates are negligible compared to the stresses needed to produce appreciable deformation. The dimensionless group σ/G governs the extent of deformation. Only for relatively soft particles suspended in a highly viscous medium would it be feasible to produce shear stresses sufficiently high to cause deformation.

3. VISCOSITY VERSUS CONCENTRATION

Figure 2 shows the concentration dependency of the relative viscosity of a typical polymer colloid. The intrinsic viscosity gives its initial

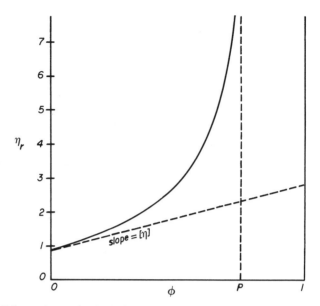

Fig. 2. Schematic graph of relative viscosity as a function of volume fraction.

slope, and the relative viscosity approaches infinity as the volume fraction approaches the maximum concentration ϕ_m at which the system can still flow. According to Maclaurin's theorem, therefore, the functional dependency of η_r on ϕ can be written as:

$$\eta_r = 1 + [\eta]\phi + b_2\phi^2 + b_3\phi^3 \ldots \tag{15}$$

The coefficient b_i is then given by the ith derivative of the function, evaluated at $\phi = 0$ and divided by $i!$. Theoretical calculation of these derivatives, even in the simplest case of rigid spherical particles, is a formidable task. To evaluate b_i, one must consider simultaneous hydrodynamic interactions among i particles. The only successful treatment to date is that of Batchelor and Greene[10] for b_2. Their calculations, which apply to extensional rather than shear flows, give $b_2 = 7\cdot6$. There appears to be little hope that the higher coefficients will be soon forthcoming, and the difficulties in calculating b for shear flows appear insurmountable.

Truncation of an infinite power series such as that given by eqn (15) does not appear to be a useful approach, even empirically. When experimental data are fitted by such a series, the values of the coefficients depend appreciably on where the series is truncated. And, taken literally, a truncated series predicts finite viscosity even when $\phi > 1$.

A purely empirical approach to dependency of η_r on ϕ would seek suitable functions in which adjustable parameters are embedded, and then find the best fit obtainable through variation of the parameters. In 1962, Rutgers[11] compiled a list of over 100 functions which had been proposed in the scientific literature to describe the dependency of η_r on concentration. He concluded that most of them are intrinsically unsuitable.

An approach which has been very useful considers the final suspension to have been constituted by two or more successive additions of particles to the suspending medium. At each addition, the suspension previously formed is considered to act as a continuum towards the new increment of particles. This leads to a functional equation:

$$\eta_r(\phi_1 + \phi_2) = \eta_r(\phi_1)\eta_r(\phi_2) \tag{16}$$

The only solution giving the correct intrinsic viscosity is due to Arrhenius:[12]

$$\eta_r = \exp[\eta]\phi \tag{17}$$

Equation (17) fits quite well at low concentrations, but fails at high concentrations, predicting finite viscosity for $\phi > 1$. A modification of this approach by Mooney[13] gives a result which fits rigid-sphere data up to values of ϕ of about $0·35$. Mooney pointed out that, when the second increment of particles is added, not all of the dispersion is available to the new particles. A proportion $k\phi_1$ of the volume is inaccessible to the new particles, where k is a co-volume factor very much like that in van der Waals' theory of gases. Similarly, a fraction $k\phi_2$ of the volume is inaccessible to the first increment of particles, due to the volume pre-empted by the second increment. Mooney's functional equation is then:

$$\eta_r(\phi_1 + \phi_2) = \eta_r\left(\frac{\phi_1}{1 - k\phi_2}\right)\eta_r\left(\frac{\phi_2}{1 - k\phi_1}\right) \tag{18}$$

The solution of eqn (18) is:

$$\eta_r = \exp\left(\frac{[\eta]\phi}{1 - k\phi}\right) = \exp\left(\frac{[\eta]\phi}{1 - \phi/\phi_m}\right) \tag{19}$$

where $\phi_m = 1/k$ can be looked upon as the maximum concentration where flow is possible.

Mooney's approach, with the co-volume taken into account in both factors, over-corrects at high concentrations. Dougherty[14] included the co-volume only once, giving the functional equation:

$$\eta_r(\phi_1 + \phi_2) = \eta_r(\phi_1)\eta_r\left(\frac{\phi_2}{1 - k\phi_1}\right) = \eta_r(\phi_2)\eta_r\left(\frac{\phi_2}{1 - k\phi_1}\right) \qquad (20)$$

The solution to this functional equation is:

$$\eta_r = \left(1 - \frac{\phi}{\phi_m}\right)^{-[\eta]\phi_m} \qquad (21)$$

Equation (21) fits rigid-sphere data to above $\phi = 0.50$, with reasonable values of the parameters $[\eta]$ and ϕ_m. An equally good fit can be obtained with a modification of Eilers'[15] empirical equation:

$$\eta_r = \left\{1 + \frac{[\eta]\phi}{2(1 - \phi/\phi_m)}\right\}^2 \qquad (22)$$

All three of the equations cited above contain the same two parameters: $[\eta]$ and ϕ_m. Theoretically, $[\eta]$ should be 2.5 for rigid spheres, and ϕ_m should be about 0.62 if the spheres are of uniform diameter. Values close to these have been obtained by fitting eqns (21) and (22). Mooney's equation, when fitted to data for $\phi > 0.35$, usually gives unrealistically high values for ϕ_m. Obviously, equations containing more adjustable parameters will be capable of better representing experimental data.

Dispersions containing a range of particle sizes are said to be polydisperse. It is easy to visualise a bimodal system which would flow at volume fractions in excess of 74%, the concentration of close-packed uniform spheres. All that is needed is for the smaller particles to fit within the voids of the array of large spheres. Mooney's analysis included polydisperse systems, and Farris[16] devised a scheme to derive the viscosity–concentration behaviour of a polydisperse system of known distribution from the corresponding function for the monodisperse system. When one of the two-parameter models is used, the packing fraction would be expected to increase as the system becomes more polydisperse.

Several specialised approaches have been developed to treat highly concentrated dispersions. Simha[17] and others have used a cage model, while Frankel and Acrivos[18] favour an approach based on lubrication theory. Neither of these approaches applies directly to the dilute range. Prager's[19] method, based on pair correlation function and a minimum dissipation theorem, has been successful only in providing widely separated upper and lower bounds to the relative viscosity.

4. NON-NEWTONIAN VISCOSITY

At each concentration, the relative viscosity can be expressed as a function of the shear stress. Polymer colloids are Newtonian at low volume fractions, and become increasingly non-Newtonian as concentration increases. It may be possible to represent this behaviour by means of a viscosity–concentration relationship whose parameters depend functionally on the shear stress. In the two-parameter model, the packing fraction would bear the entire burden of expressing shear dependence, since the intrinsic viscosity of a rigid-sphere dispersion is independent of shear. A more common way to describe non-Newtonian behaviour is to fit the same viscosity–shear stress equation to the data at each concentration, and then to develop the functional dependence of its parameters on volume fraction.

The first widely used non-Newtonian equation was proposed by Bingham.[20] The 'Bingham plastic' model shows no flow at stresses below a 'yield point' σ_0; at larger stresses, the shear rate is linear with stress.

$$\dot{\gamma} = \frac{\sigma - \sigma_0}{\eta_1} \qquad (23)$$

Here η_1 represents the limiting viscosity at high shear rates. At shear stresses above the yield stress, Casson's[21] modification of Bingham's model has the square root of the shear rate depending linearly on the square root of the shear stress:

$$\dot{\gamma}^{\frac{1}{2}} = \frac{\sigma^{\frac{1}{2}} - \sigma_0^{\frac{1}{2}}}{\eta_1^{\frac{1}{2}}} \qquad (24)$$

Casson's model has been applied with considerable success to those polymer colloids that exhibit the yielding type of behaviour.

Krieger and Maron[22] showed that synthetic rubber latex flows even at very low stresses, and that the shear dependence of the viscosity can

be represented over a moderate range of shear rates by the Ostwald–
de Waele 'power law':

$$\dot{\gamma} = a\sigma^n \tag{25}$$

Its parameters are a coefficient a, which is numerically equal to the
viscosity at unit shear stress, and an exponent n.
Several theoretical models have been developed to describe shear-
thinning non-Newtonian dispersions. Ree and Eyring[23] applied reaction rate theory to the physical process of shear flow. Maron and
Pierce[24] adapted the Ree–Eyring equation to polymer colloids by
considering only two types of flow unit, a non-Newtonian component
corresponding to the particle and a Newtonian unit corresponding to
the medium:

$$\eta = \eta_1 + (\eta_2 - \eta_1)\frac{\sinh^{-1}(\beta\dot{\gamma})}{\beta\dot{\gamma}} \tag{26}$$

This equation predicts Newtonian regimes at both high and low shear
stresses, as is often observed with polymer colloids. η_2 is thus the
low-shear limiting viscosity, and β is a relaxation time. Similar behaviour is shown by an equation derived by Dougherty and
Krieger,[14,25] based on proximity doublets which separate under shear:

$$\eta = \eta_1 + \frac{\eta_2 - \eta_1}{1 + |\sigma|/\sigma_c} \tag{27}$$

where σ_c is a characteristic shear stress, proportional to kT/a^3. Equation (27) was fitted by Papir and Krieger[26] to data on uniform polymer
colloids of different particle sizes dispersed in media of different
viscosities. To express the concentration dependence of the limiting
viscosities, they used eqn (21) with $[\eta] = 2.67$ and two different values
for the packing fraction: 0.57 for the low-shear limit and 0.68 at high
shear.
Cross's[27] equation is similar to eqn (27) above, but contains an
additional parameter n.

$$\eta = \eta_1 + \frac{\eta_2 - \eta_1}{1 + (\dot{\gamma}/\dot{\gamma}_c)^n} \tag{28}$$

This parameter usually takes on values between $\frac{2}{3}$ and 1 when the
equation is fitted to experimental data.
There are thus two different types of flow behaviour represented by
the non-Newtonian flow equations above. Equations (23) and (24)

describe systems which are solid-like at stresses below a yield stress, while eqns (26) to (28) describe systems which show a finite Newtonian viscosity at low stresses. (Equation (25) should be regarded as an interpolation formula useful only over limited ranges of shear stress.) These apparently exclusive types of flow behaviour are reconciled in recent studies by Buscall et al.[28] on polymer colloids with strong electrostatic interactions among the particles. They used a commercial rotational rheometer, which can exert very small shear stresses and measure the resulting very small shear rates, to study systems which appear to show yield stresses. They found that, at stresses below the yield stress, these systems flowed as Newtonian liquids of very high viscosity.

Shear-thinning non-Newtonian fluids apparently undergo some form of structural transition; indeed, the term 'structural viscosity' has been applied to this type of flow behaviour. Theories of shear-thinning behaviour postulate that the dispersion at rest takes on a structure of short or long range. This structure is broken down by shearing forces, so that the high-shear limit is a dispersion of individual particles. At each steady shear rate, the prevailing viscosity is attributable to the degree of structure which expresses a balance between the attractive or repulsive forces which produce structure and the shearing forces which destroy it.[29]

When a shear-thinning fluid is subjected to a transient jump in shear rate, the system will relax to that degree of structure which corresponds to the new shear rate. If the time for this relaxation is long enough, it may be possible to follow the process by observing the time dependence of the viscosity. The term 'thixotropy', originally coined by Freundlich[30] to describe a reversible sol–gel transformation, has been applied to the phenomenon of shear-thinning in general, and especially to its time-dependent aspects. The structural relaxation times of most polymer colloids are so small that time dependence is difficult to observe. Polymer colloids thickened by addition of water-soluble polymer show longer relaxation times, as shown by Mooney.[31]

Shear-thickening behaviour has been observed in polymer colloids at high shear rates. In a classical study combining optical diffraction with rheological measurements, Hoffman[32] observed a uniform polymer colloid in shear flow. He found that the particles organised into two-dimensional hexagonal arrays normal to the velocity gradient. At higher shear rates these arrays became unstable and buckled, leading to a sharp increase in the measured viscosity. Choi[33] observed shear-thickening behaviour in a sterically stabilised polymer colloid, but

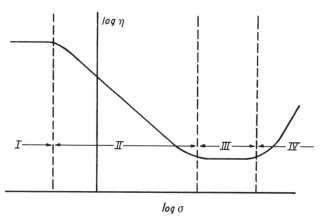

Fig. 3. Schematic graph of the four flow regimes: (I) Newtonian flow of randomly ordered dispersion; (II) shear thinning as layers develop; (III) fully layered flow; (IV) shear thickening due to instability of layered flow.

found the transition to the shear-thickening regime in his system to be a gradual one.

Hoffman[32] has proposed a flow mechanism, based on the formation of layered arrays, which covers the entire shear-rate range. As described by Choi in Fig. 3, there are four sequential regimes of flow. At low stresses, the system displays some form of three-dimensional order (Region I). As shear stress increases, the system gradually transforms into a structure of two-dimensional hexagonal layers. Region II is the shear-thinning transition regime, while Region III represents the layered structure whose viscosity is that of the high-shear Newtonian regime. Region IV, the shear-thickening regime, results from instability of the layered structure at very high shear rates. Hoffman's explanation of this instability involves interparticle forces, while Choi's[33] explanation is based on competition between rates of shear flow and of diffusion.

5. DIMENSIONAL ANALYSIS[34]

The viscosity η of a dispersion of non-interacting uniform spheres in a Newtonian medium can depend on as many as eight independent variables. These include the density ρ_0 and viscosity η_0 of the medium, the density ρ, radius a and concentration N (number per unit volume)

TABLE 1
Variables in the Rheological Equation for Rigid-Sphere Dispersions (M = mass, L = length, T = time)

Variable	Symbol	Dimensions
Viscosity of dispersion	η	$ML^{-1}T^{-1}$
Viscosity of medium	η_0	$ML^{-1}T^{-1}$
Density of medium	ρ_0	ML^{-3}
Density of particle	ρ_s	ML^{-3}
Radius of particle	a	L
Number density of particles	N	L^{-3}
Rate of shear	$\dot{\gamma}$	T^{-1}
Elapsed time	t	T
Thermal energy	kT	ML^2T^{-2}

of the particles, and additional parameters such as temperature T, shear rate $\dot{\gamma}$ and time t of shearing. This makes the governing equation an equation in the nine variables listed in Table 1. It is prudent to take advantage of the principle of dimensional analysis to reduce the number of variables.

Dimensional analysis treats an equation in V variables, all of which can be expressed in terms of U dimensional units (such as mass, length and time). It states that this equation can be reformulated in terms of $D = V - U$ independent dimensionless variables. (Independence means that none of the D dimensionless variables can be expressed as a product of the other $D - 1$ dimensionless variables.) Each of the new dimensionless variables is a product of the original V variables, each raised to some power (positive, negative or zero). In this case, with the temperature variable expressed as kT, all nine variables can be expressed in units of mass, length and time, and hence $U = 3$. Thus $D = 9 - 3 = 6$.

Table 2 presents a list of six independent dimensionless variables derived from the original nine. Other dimensionless variables can be constructed as products of these six, but these would not meet the criterion of independence. They may, however, prove convenient in describing certain experimental situations. Among the six dimensionless variables is the dimensionless concentration ϕ and the dimensionless viscosity η_r. A dimensionless shear stress is used in preference to a dimensionless shear rate; their ratio would be the relative viscosity. For slow (Re $\rightarrow 0$) steady ($t_r \rightarrow \infty$) flow of a dispersion of neutrally

TABLE 2
Dimensionless Groups in Rheology of Rigid-Sphere Dispersions

Group	Symbol	Name
$\dfrac{\eta}{\eta_0}$	η_r	Relative viscosity
$\dfrac{4\pi Na^3}{3}$	ϕ	Volume fraction
$\dfrac{\sigma a^3}{kT}$	σ_r	Reduced shear stress
$\dfrac{\rho_s}{\rho_0}$	ρ_r	Relative density
$\dfrac{kTt}{\eta_0 a^3}$	t_r	Reduced time
$\dfrac{a^2\dot{\gamma}\rho_0}{\eta_0}$	Re_i	Internal Reynolds number

buoyant ($\rho_r \rightarrow 1$) particles, the problem becomes one in three dimensionless variables:

$$\eta_r = f(\phi, \sigma_r) \tag{29}$$

Experimental definition of the functional dependency expressed by eqn (29) is a tractable problem.

Chaffey[35] has outlined extensions of the principle of dimensional analysis to more complex dispersions, involving particle interactions and deformations.

6. NON-INTERACTING PARTICLES

Woods and Krieger[36] prepared a series of five uniform poly(styrene) latices ranging in particle diameter from 220 to 830 μm. The concentration of the stabilising surfactant was adjusted to just below monolayer coverage on the particles, as determined from surface tension measurements. Viscosities were measured as functions of added electrolyte, in order to determine the amount needed to provide an ionic atmosphere which would minimise particle–particle interactions. They also determined the effective thickness of the adsorbed surfactant layer, and included it as part of the effective volume fraction. Viscosity measurements were made as functions of shear stress, at volume fractions ranging up to 55%.

When relative viscosities were plotted against the dimensionless shear stress, data for the three particle sizes below 500 nm superimposed, while the two larger particle sizes fell on separate curves. Figure 4 shows the data for $\phi = 0.50$. Failure of particles larger than 500 nm to follow the predictions of the dimensional analysis was attributed to the importance in large-particle systems of London–van der Waals forces, which were ignored in the analysis.

In a subsequent study, Papir and Krieger[37] redispersed crosslinked uniform poly(styrene) latex particles of four different particle sizes into benzyl alcohol ($\eta_0 = 0.05$ cP) and into m-cresol ($\eta_0 = 0.17$ cP). Figures 5(a) and 5(b) show the viscosities of the respective 50% volume fraction dispersions as functions of the shear stress. Figure 6 shows

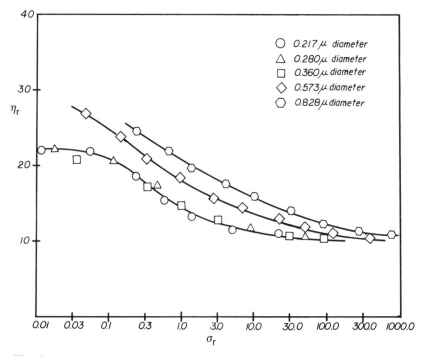

Fig. 4. Relative viscosity versus reduced shear stress for monodisperse latices of five different particle sizes at $\phi = 0.50$.

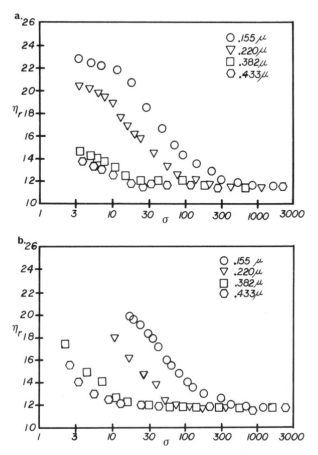

Fig. 5. Relative viscosity versus shear stress for four different particle sizes at 0·50 volume fraction in: (a) benzyl alcohol, and (b) m-cresol.

both sets of data as functions of the dimensionless shear stress, together with the curve for the 50% aqueous latices taken from Fig. 4. The superposition of the three sets of data thus covers a three-fold range of particle size and a 20-fold range of viscosity of the medium.

7. THE SECOND ELECTROVISCOUS EFFECT

The second electroviscous effect, in the nomenclature of Conway and Dobry-Duclaux,[38] is the augmentation of the viscosity of a dispersion

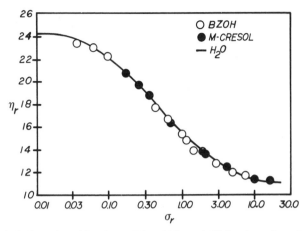

Fig. 6. Relative viscosities from Figs 5(a) and 5(b), plotted as functions of reduced shear stress. Solid line represents data on aqueous latices taken from Fig. 4.

due to particle–particle electrostatic repulsion. Fryling[39] and Brodnyan and Kelley[40] studied the second electroviscous effect in dialysed synthetic rubber latices. Krieger and Eguiluz[41] prepared uniform polystyrene latices, and subjected them to a mixed-bed ion exchange resin to remove excess electrolyte. The open circles in Fig. 7 show their relative viscosity versus dimensionless shear stress curve for the deionised latex at 40% volume fraction. The other curves show the effect of added electrolyte (HCl) in reducing the viscosity.

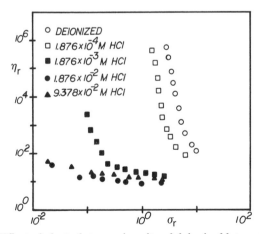

Fig. 7. Effect of electrolyte on viscosity of deionised latex, $\phi = 0 \cdot 40$.

The curves at low levels of added electrolyte appear to exhibit yield stresses, and all the curves appear to approach asymptotically the same high-shear viscosity. These impressions are confirmed by plotting the data as the square root of reduced shear stress against the square root of the reduced shear rate, as in Fig. 8. Linearity of these plots indicates that Casson's[21] equation is followed; the intercepts of the lines give the square root of the yield stress. Parallelism of the lines indicates the same value of η_1, the high-shear limiting viscosity.

A second aspect of the study compared the effects of electrolytes of different valency. An unexpected result is that the reduction in viscosity is the same for all electrolytes at the same equivalent concentration (i.e. at the same normality). This is exemplified in Fig. 9, which compares the effects of K_2SO_4 and $Th(NO_3)_4$.

The work of Krieger and Eguiluz[41] seems to show that deionised latices set up a solid-like structure, which is broken down by both shear and electrolyte addition. The crystal-like order of these systems gives rise to a colourful iridescence, due to Bragg diffraction of visible light. This phenomenon has itself been the subject of a large number of studies.[42–44] Both the shear elasticity and the viscosity of such polymer colloids have been measured by researchers at the University of

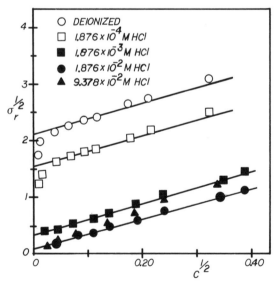

Fig. 8. Data of Fig. 7, plotted to test applicability of Casson's equation.

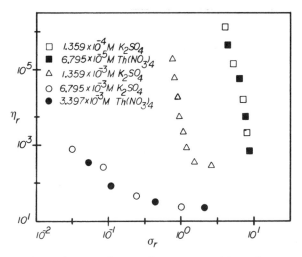

Fig. 9. Comparison of electroviscous effects produced by K_2SO_4 and $Th(NO_3)_4$ at the same normalities.

Bristol.[28,45] They showed that the shear modulus is insensitive to electrolyte level until the concentration of added electrolyte exceeds the concentration of the counterions. The study of Buscall et al.[28] demonstrated flow at stresses below the yield stress. The low-shear limiting viscosities they found were extremely high, and their order of magnitude could be estimated from electrostatic interaction energies via the Ree–Eyring theory as shown by Goodwin et al.[45]

Benzing and Russel[46] have developed a self-consistent field theory of the second electroviscous effect, and have compared its predictions with experimental measurements, with encouraging results.

8. STERICALLY STABILISED DISPERSIONS

For electrostatically stabilised dispersions, the range of interparticle repulsions can be quite long, conferring long-time stability against coagulation. Many polymer colloids are stabilised wholly or in part by bound or adsorbed polymer fragments. When the surfaces of two particles are separated by distances less than the sum of the thicknesses of their stabilising polymer layers, a strong repulsive force will arise. This force is variously ascribed to steric interference and to osmotic effects. When, however, the separation of the particle surfaces exceeds the sum of the layer thicknesses, there is no repulsion arising from this source. Nevertheless, the strong short-range repulsion arising from steric stabilisation gives the dispersion thermodynamic stability. Figures 10(a) and 10(b) are schematics which compare potential energy versus distance for electrostatic and steric stabilisation, respectively. In a study of sterically stabilised polyvinyl chloride dispersions of various particle sizes in organic media, Willey and Macosko[47] achieved superposition by plotting relative viscosity against a reduced shear rate. However, their relative viscosities were considerably higher than those of Fig. 6, due perhaps to swelling of the particles.

Because the thickness d of the polymer layer is often an appreciable fraction of the particle radius a, the polymer layer augments significantly the hydrodynamic volume of the particle. The effective volume fraction ϕ of the colloid is therefore greater than the volume fraction ϕ_0 of the particles themselves:

$$\phi = \phi_0 \left[1 + \left(\frac{d}{a} \right)^3 \right] \tag{30}$$

The effect of the stabilising layer is well documented in the study by Choi.[33] Following Everett and Stageman,[48] he polymerised poly(methyl methacrylate) in hexane, using as stabiliser a triblock copolymer with end blocks of poly(dimethyl siloxane) (PDMS) and a central block of poly(styrene) (PS). By varying the proportion of stabiliser, he obtained uniform spherical particles of radii ranging from 95 to 315 nm. He then redispersed the particles into a series of silicone fluids ranging in nominal viscosities from 10 to 200 cSt. Volume fractions ranged from 10 to 40%. He found that an assumed layer thickness of 7 nm for all particle sizes gave intrinsic viscosities of 2·5, and also brought the viscosity–concentration data into internal consistency. This layer thickness was consistent with the difference between particle radii measured by electron microscopy and by photon correlation spectroscopy.

Measuring steady-shear viscosities of his dispersions at 30°C, Choi found two quite distinct types of behaviour, designated here as fluid and gelling. Figure 11 shows the two types of behaviour, as observed with particles 20·4 nm in diameter at $\phi = 21\%$. All dispersions in the 10-cSt silicone flowed easily even at negligible shear stress, and, when compared at the same effective volume fractions, their relative viscosity versus reduced shear stress curves superposed (Fig. 12). On the

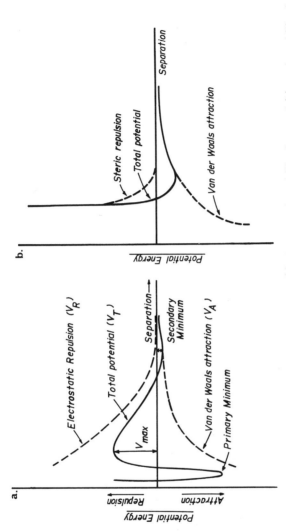

Fig. 10. Comparison of interparticle potentials due to: (a) electrostatic stabilisation, and (b) steric stabilisation.

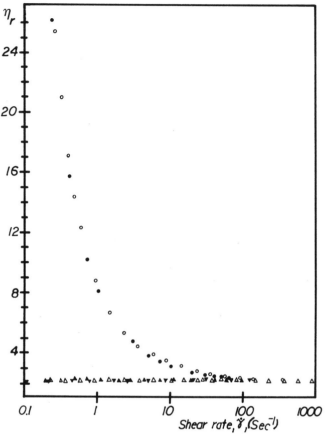

Fig. 11. Comparison of fluid and gelling dispersions at $\phi = 0.21$. Particle diameter, 410 nm. 'Gelling' dispersion, silicone MW = 11 000; $\eta_0 = 1.81$ P at $T = 30.0°C$: $\bigcirc$, Weissenberg rheogoniometer; $\bullet$, mechanical spectrometer. 'Fluid' dispersion, silicone MW = 4400; $\triangle$, $T = 30.0°C$, $\eta_0 = 0.441$ P; $\blacktriangle$, $T = -21.0°C$, $\eta_0 = 1.70$ P; $\blacktriangledown$, $T = -44.0°C$, $\eta_0 - 4.46$ P.

other hand, all dispersions in the 200-cSt fluid showed gel-like behaviour (Fig. 13). They exhibited apparent yield stresses, and their non-Newtonian viscosities followed Casson's[21] equation fairly closely (Fig. 14). High-shear limiting relative viscosities were the same as those for the fluid dispersions at the same effective volume fractions. Dispersed in 50-cSt silicone, the larger particles gave fluid dispersions, while the smallest particles gave gelling dispersions.

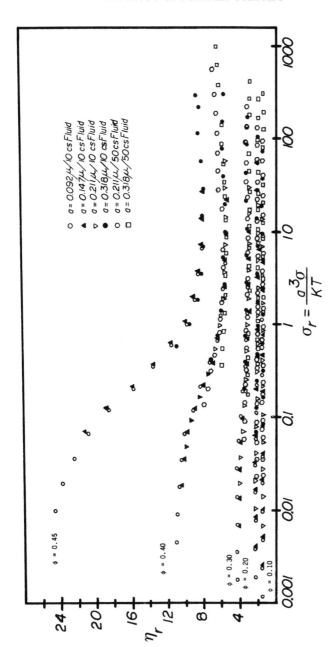

Fig. 12. Superposition of relative viscosities versus reduced shear stresses for 'fluid' sterically stabilised polymer colloids. Data represent four different particle sizes in 10-cSt fluid and the two largest particle sizes in 50-cSt fluid.

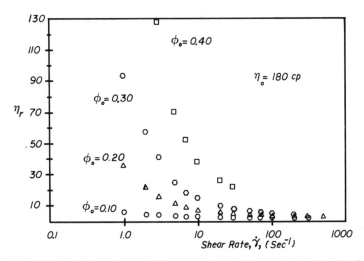

Fig. 13. 'Gelling' dispersions: 623-nm particles dispersed in 200-cSt fluid; $T = 30°C$.

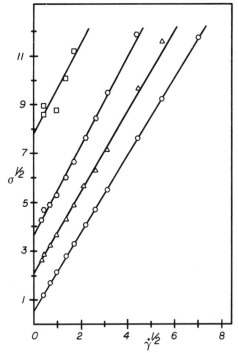

Fig. 14. Data from Fig. 13 plotted to test applicability of Casson's equation.

In order to investigate the role of the viscosity of the silicone medium in determining whether the dispersion is fluid or gelling, Choi made rheological measurements at temperatures ranging down to −44°C, where the '10-cSt' fluid has a viscosity exceeding 400 cSt. He found that those dispersions which were fluid at 30°C remained fluid at even the lowest temperatures, and that their relative viscosity versus reduced shear stress curves superposed at all temperatures (Fig. 11). This points to the molecular weight of the medium, rather than its viscosity, as the determining factor.

Gelation implies the existence of a three-dimensional network structure. Since the 200-cSt fluid has a molecular weight of 11 000, comparable to that of the PDMS block of the stabiliser, such a structure could be attributed either to bridging or to steric interference. For bridging to occur, the silicone molecules of the medium would have to entangle with the stabilising fragments of adjacent particles. Their molecular weights, however, are too low to produce entanglement. Steric interference would cause a repulsive potential between particles separated by distances greater than the sum of the thicknesses of their stabilising layers. Two experimental observations point to a repulsive potential. One is the resemblance of the rheological behaviour of the gelling dispersions to that of aqueous dispersions under the second elec-

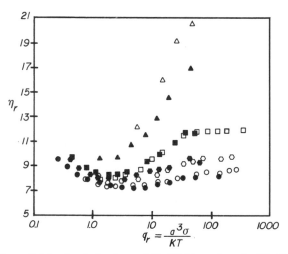

Fig. 15. Shear-thickening behaviour of 'fluid' dispersions in 50-cSt fluid. Open points represent 420-nm particles; closed points represent 640-nm particles. △▲, 446 cP $(T = -44 \cdot 0°C)$; □■, 170 cP $(T = -21 \cdot 0°C)$; ◯◆, 91 cP $(T = 0°C)$; ◯●, 44·1 cP $(T = 30 \cdot 0°C)$.

troviscous effect. The other is the fact that viscosities of the gelling dispersions increase with temperature.

As described earlier in this chapter, Choi[33] found that the fluid dispersions exhibited shear-thickening behaviour at high shear rates. As shown in Fig. 15, shear-thickening sets in smoothly at reduced shear stresses in the range 1–10, and is much more pronounced at lower temperatures where the medium is more viscous.

9. SOME UNSOLVED PROBLEMS

Most of the experimental studies of the rheology of polymer colloids have focussed on systems of uniform particle size. Multimodal and broad distributions will be less viscous than uniform distributions of the same volume fraction. This is of great practical interest since fluid systems of high concentration are desirable for efficiency in synthesis, formulation and application of polymer products.

Polymers soluble in the medium are often present in polymer colloids, sometimes as part of the polymerisation recipe or as by-products of polymerisation. More often, however, they are added to modify rheological properties of the dispersion. Some adsorption onto the particle surface can be expected, so that the effect of soluble polymer is not solely an increase in the viscosity of the medium. Interaction between adsorbed and dissolved polymers may cause gelation, through bridging or through steric interference. Both experimental work and theoretical work are needed to elucidate the role of soluble polymers in the rheology of polymer colloid systems.

The layered-flow theory advanced by Hoffman[32] and by Choi[33] appears very attractive, but as yet it is mainly qualitative. Two pieces of quantitative theoretical work are needed to confirm or disprove the theory. The first is to derive the concentration dependency of the high-shear limiting relative viscosity on the basis of the layered-flow model. The second is to calculate for the same model the viscosity as a function of shear stress in the shear-thinning region.

The rheological effect of interparticle forces has been reviewed by Russel.[49] He points to some successes in dealing with ordered polymer colloids, but to less satisfactory situations for dispersions which are disordered or flocculated.

The problems which remain unsolved are numerous and difficult. The pace and quality of recent work, however, gives reason to expect much progress in the years to come.

REFERENCES

1. Einstein, A., *Ann. Physik*, **17**, 459 (1905); **19**, 271 (1906); **34**, 591 (1911).
2. Frisch, H. L. and Simha, R., Ch 14 in *Rheology: theory and applications*, Vol. 1, F. R. Eirich (ed.), Academic Press, New York (1936).
3. Simha, R., J. *Phys. Chem.*, **44**, 25 (1940); J. *Chem. Phys.*, **13**, 188 (1945).
4. Kuhn, W. and Kuhn, H., *Helv. Chim. Acta*, **28**, 97 (1945).
5. Eirich, F. R., Bunzl, M. and Margaretha, H., *Kolloid-Z.*, **74**, 276 (1936).
6. Saunders, F. L., J. *Colloid Sci.*, **16**, 13 (1961).
7. von Smoluchowski, M., *Kolloid-Z.*, **18**, 190 (1916).
8. Booth, F., *Proc. Roy. Soc.*, **A203**, 533 (1950).
9. McDonogh, R. W. and Hunter, R. J., J. *Rheology*, **27**, 189 (1983).
10. Batchelor, G. K. and Greene, J. T., J. *Fluid Mech.*, **56**, 401 (1972).
11. Rutgers, I. R., *Rheol. Acta*, **2**, 305 (1962).
12. Arrhenius, S., Z. *Physik. Chem. (Leipzig)*, **1**, 285 (1887).
13. Mooney, M., J. *Colloid Sci.*, **6**, 162 (1951).
14. Dougherty, T. J., PhD thesis, Case Institute of Technology (1959).
15. Eilers, H., *Kolloid-Z.*, **97**, 913 (1941); **102**, 154 (1943).
16. Farris, R. J., *Trans. Soc. Rheology*, **12**, 281 (1968).
17. Simha, R., J. *Appl. Phys.*, **23**, 1020 (1952).
18. Frankel, N. A. and Acrivos, A., *Chem. Eng. Sci.*, **22**, 847 (1967).
19. Prager, S., *Physica*, **29**, 129 (1963).
20. Bingham, E. C., US Bureau of Standards Sci. Paper No. 278 (1916).
21. Casson, N., In *Rheology of disperse systems*, C. C. Mill (ed.), Pergamon Press, New York (1959).
22. Krieger, I. M. and Maron, S. H., J. *Colloid Sci.*, **6**, 528 (1951).
23. Ree, T. and Eyring, H., J. *Appl. Phys.*, **26**, 793, 800 (1955).
24. Maron, S. H. and Pierce, P. E., J. *Colloid Sci.*, **11**, 80 (1956).
25. Krieger, I. M. and Dougherty, T. J., *Trans. Soc. Rheology*, **3**, 137 (1959).
26. Papir, Y. S. and Krieger, I. M., J. *Colloid & Interface Sci.*, **34**, 126 (1970).
27. Cross, M. M., J. *Colloid Sci.*, **20**, 414 (1965).
28. Buscall, R., Goodwin, J. W., Hawkins, M. W. and Ottewill, R. H., J. *Chem. Soc., Faraday Trans. I*, **78** (10), 2873, 2889 (1982).
29. Cheng, D. C. H., J. *Phys.*, **E4**, 693 (1971).
30. Freundlich, H., *Kapillarchemie*, 3rd edn., Leipzig (1932); *Thixotropy*, Hermann, Paris (1935).
31. Mooney, M., J. *Colloid Sci.*, **1**, 195 (1946).
32. Hoffman, R. L., J. *Colloid & Interface Sci.*, **46**, 491 (1974).
33. Choi, G. N., PhD thesis, Case Western Reserve University (1982).
34. Krieger, I. M., *Trans. Soc. Rheology*, **7**, 101 (1963).
35. Chaffey, C. E., *Colloid & Polymer Sci.*, **255**, 691 (1977).
36. Woods, M. E. and Krieger, I. M., J. *Colloid & Interface Sci.*, **34**, 91 (1970).
37. Papir, Y. S. and Krieger, I. M., J. *Colloid & Interface Sci.*, **34**, 126 (1970).
38. Conway, B. E. and Dobry-Duclaux, A., Ch 3 in *Rheology: theory and applications*, Vol. 3, F. R. Eirich (ed.), Academic Press, New York (1960).
39. Fryling, C. F., J. *Colloid Sci.*, **18**, 713 (1963).
40. Brodnyan, J. G. and Kelley, E. L., J. *Colloid Sci.*, **19**, 488 (1964).
41. Krieger, I. M. and Eguiluz, M., *Trans. Soc. Rheology*, **20**, 29 (1976).

42. Hiltner, P. A. and Krieger, I. M., *J. Phys. Chem.*, **73**, 2386 (1969).
43. Hachisu, S., Kobayashi, Y. and Kose, A., *J. Colloid & Interface Sci.*, **42**, 342 (1973).
44. van Megen, W. and Snook, I., *J. Colloid & Interface Sci.*, **57**, 40 (1976).
45. Goodwin, J. W., Gregory, T. and Stile, J. A., *Adv. Colloid Interface Sci.*, **17**, 185 (1982).
46. Benzing, D. W. and Russel, W. B., *J. Colloid & Interface Sci.*, **83**, 167, 178 (1981).
47. Willey, S. J. and Macosko, C. W., *J. Rheology*, **20**, 525 (1978).
48. Everett, D. H. and Stageman, J. F., *Faraday Disc. Chemical Society*, **65**, 230 (1978).
49. Russel, W. B., Publ. Math. Res. Center, Univ. Wisconsin–Madison **49** (1983).

Chapter 7

Natural and Synthetic Rubber Latices

D. C. BLACKLEY

*London School of Polymer Technology,
The Polytechnic of North London, London, UK*

1. INTRODUCTION

The purpose of this chapter is to review briefly certain aspects of the science and technology of rubber latices. Because of limitations of space, the treatment here must necessarily be superficial and incomplete. A two-volume work comprising some 850 pages on this subject was published by the present author in 1966;[1] not only was that treatment incomplete in 1966, but it is now, of course, seriously out of date. Certain aspects of the subject have been reviewed in a recent book which has been edited by Calvert.[2] In the present chapter, it is possible to do no more than introduce the reader to the more important facets of the subject, and to indicate where more detailed information can be obtained if it is required.

For the purposes of this chapter, it is convenient to define a *latex* as *a stable dispersion of a polymeric substance in an essentially aqueous medium*. It is to be understood that the polymer particles, which are usually approximately spherical in shape, are of colloidal dimensions, say, between 30 and 500 nm in diameter, and that the dispersion is relatively concentrated, the volume fraction of contained polymer being within the range 0·40–0·70. It is also to be understood that the dispersion is stable relative to the flocculated or coagulated state, not in the absolute thermodynamic sense, but in the kinetic sense usually understood for lyophobic colloids generally. Thus the ability to persist as a colloidal dispersion over a period of months or years is a consequence of the slow rate at which aggregation of the particles

247

occurs, rather than of absolute thermodynamic stability. Concerning the nature of the dispersion medium, although some interest has been shown in recent years in the production and application of dispersions of polymer particles of colloidal dimensions in non-aqueous media, it is the case that the majority of latices used industrially have aqueous dispersion media, and so conform to the above definition. Certainly as far as the latices of rubbery polymers are concerned, there seems to have been little or no serious interest in products which have non-aqueous dispersion media.

The subject of this chapter is *rubber* latices. The feature of this sub-group of polymer latices which distinguishes them from latices in general is that the contained polymer is rubbery—and therefore above its glass-transition temperature—at room temperature. This feature has several practical consequences, of which the more obvious and important are as follows:

(1) The particles will deform more readily than will the particles of a non-rubber latex when they collide as a consequence of Brownian motion.

(2) The particles will integrate more readily to form a coherent film when the latex is dried down than will the particles in a non-rubber latex.

(3) Molecules adsorbed at the interface between the particles and the dispersion medium are probably more mobile than are similar molecules adsorbed on the surface of non-rubbery polymer particles, and this may have important consequences in respect of the colloid stability imparted by the presence of those adsorbed molecules.

Apart from any classification on the basis of the chemical nature of the contained polymer, rubber latices are conveniently classified generally into three groups as follows:

(1) *natural latices,* which are those which are produced naturally by metabolic processes which occur in certain species of plant

(2) *synthetic latices,* which are produced by the emulsion polymerisation of appropriate monomers using an aqueous dispersion medium

(3) *artificial latices,* which are produced by dispersing rubbers in bulk form in an aqueous medium

This chapter will be principally concerned with natural and synthetic rubber latices, because these are the types which are almost exclusively of industrial importance at the present time. However, for the sake of completeness, and because they may eventually become important industrially, brief mention will also be made of artificial rubber latices. It is appropriate to conclude this introduction by listing the many and varied characteristics of a rubber latex which ideally require to be specified in order to define a rubber latex fairly completely. For most rubber latices, they fall conveniently under three headings:

(1) those which pertain to the *contained polymer*
(2) those which pertain to the *aqueous dispersion medium*
(3) those which pertain to the *interfacial region* between the particles and the dispersion medium

Those characteristics which pertain to the contained polymer fall into two main groups:

(1) those which relate to what may be described as the colloidal aspects of the dispersion, and include the volume fraction of contained polymer and matters such as particle shape and particle size and size distribution
(2) those which relate to the polymer itself, in which group are included matters such as the chemical nature of the polymer, the nature and concentration of any functional groups which may be present, the morphology of the polymer particles (e.g., whether they are of uniform composition or whether they are of the core-shell or gradient-copolymer type), the molecular weight of the polymer, the gel content of the polymer, the concentration of crosslinks in any polymer gel which may be present, the glass-transition temperature(s) of the polymer, the extent to which the polymer is crystalline at normal temperatures, and certain of the mechanical properties of the polymer, notably its viscoelastic behaviour

The characteristics which pertain to the aqueous dispersion medium include principally those which relate to the natures and concentrations of substances dissolved in the dispersion medium. Such substances divide conveniently into low-molecular-weight species and macro-molecular species, and in both cases the species may be ionic or non-ionic. Of particular importance for the practical applications of rubber latices are matters such as the pH and ionic strength of the dispersion medium, and the concentration of any heavy-metal ions which may be present.

The characteristics which pertain to the interfacial region between the particles and the dispersion medium include such matters as the nature and concentration of any functional groups which may be chemically bound at the surface of the particles, the extent to which such groups are ionised if they are ionisable, the nature and concentration of any adsorbed species which may be present in the interfacial region, and the extent to which water molecules are firmly bound to the surface of the particles, either because of the presence of adsorbed species or because of the nature of the polymer particle itself. Important cognate matters are the total surface density of electric charge which is present at the surface of the particles, the surface potential to which these charges give rise, and the manner in which the surface potential decays as one moves from the particle surface into the dispersion medium.

In the case of fresh natural rubber latices, it may be necessary to include a fourth group of characteristics in addition to the three main groups described above. These are characteristics which pertain to any other particulate phases which may be present. In principle, these particulate phases may contain polymers of types other than that which is present in the principal particles, but in practice these secondary particles are usually rather complex biological structures, rather than simply polymer particles.

2. COLLOID STABILITY OF RUBBER LATICES

2.1. Factors which Influence the Colloid Stability of Rubber Latices

The factors which determine the colloid stability of a rubber latex are essentially those which determine the colloid stability of any lyophobic colloid. The stability is determined by the balance which exists between attractive and repulsive forces between two particles as they approach one another closely. The attractive forces are presumed to be quantum-wave-mechanical in origin; they are usually assumed to be of the London–van der Waals type. The repulsive forces which confer colloid stability upon the latex can in principle arise from at least four distinct origins:

(1) the presence of bound electric charges at the particle surface,

and, in particular, from the presence of the associated counterion cloud in the surrounding medium

(2) the presence of adsorbed or combined macromolecular substances at the particle surface, these macromolecular substances being such that they tend to mix with the dispersion medium

(3) the presence of bound water molecules at the particle surface

(4) the presence of macromolecular substances dissolved in the dispersion medium

The types of stabilisation conferred by these four factors are known, respectively, as *electrostatic, steric, solvation* (specifically *hydration* in this case), and *exclusion stabilisation.*

It is commonly assumed that most rubber latices are stabilised predominantly by an electrostatic mechanism. Certainly it is true that the particles in most rubber latices are electrically charged. The terms *anionic latex* and *cationic latex* are used to describe, respectively, latices whose particles carry negative surface charges and those whose particles carry positive surface charges. Almost without exception, the rubber latices which find industrial application are anionic, and their aqueous phases are usually alkaline in reaction. However, periodically interest is shown in the desirability of using cationic rubber latices for certain applications; an early example is to be found in the cationic natural rubber latices investigated by Blow.[3,4] Although the presence of electric charges on the surface of the particles of many rubber latices is well established, it is not so clear to what extent the presence of these charges *per se* contributes to the colloid stability of the latex, i.e., to what extent the latex is truly electrostatically stabilised.

Not all rubber latices contain macromolecular substances which are such that they tend to mix with the dispersion medium. But if such substances are present, and if they become adsorbed at the particle surfaces, then it is probable that their presence will contribute significantly to the colloid stability of the latex. The stabilisation conferred by the presence of adsorbed and combined water-miscible macromolecular substances *per se* will be principally that of the steric type, i.e., it is primarily a consequence of the increase in the concentration of polymer segments in the aqueous phase between two particles as those particles approach one another closely. However, since such macromolecular substances are invariably hydrophilic in nature, they will probably also contribute to the stability of the latex by binding water molecules to the particle surfaces. If, as is frequently the case, the macromolecular substance is ionised, then it may also contribute to the colloid stability of the latex by electrostatic mechanism. It should also be noted that the presence of adsorbed macromolecular substances of very high molecular weight of this type can lead to severe thickening and flocculation of rubber latices because such substances are able to bridge the particles and bind them into a coherent structure. Macromolecular substances of a proteinaceous nature are present in natural rubber latex, and they undoubtedly contribute to its colloid stability. Water-miscible macromolecular substances may also be present in carboxylated rubber latices which are produced by the emulsion copolymerisation of hydrophobic monomers and unsaturated carboxylic acids. However, their contribution to the colloid stability of the latex is obscure.

Water molecules may become firmly bound to the particles of a rubber latex for various reasons. The most important of these is probably the presence of hydrophilic substances—macromolecular or otherwise—adsorbed or combined at the particle surfaces. The presence of bound electric charges at the surface will also encourage the binding of water molecules, as also may the nature of the polymer which comprises the particle. The effect of a bound water layer surrounding the particles is two-fold:

(1) It reduces to virtually zero the interfacial free energy of the surface which effectively separates the polymer particle from the dispersion medium. It therefore reduces the thermodynamic incentive to flocculation and coagulation.

(2) It provides a mechanical barrier to coagulation by preventing the polymer particles themselves from coming into close contact.

Although it is often assumed that hydration effects play an important part in the stabilisation of rubber latices, there appears to have been few fundamental studies of the importance or otherwise of solvation effects for the stability of lyophobic colloids generally. It is also difficult to distinguish between stabilisation which is a consequence of hydration of the particles *per se* and stabilisation which is a consequence of the presence of those factors which caused the particles to become hydrated. It has already been pointed out that it is difficult to distinguish steric stabilisation caused by the presence of adsorbed or combined hydrophilic macromolecular substances from hydration stabilisation caused by water associated with the macromolecular substances. Likewise, it is not easy to distinguish electrostatic stabilisation by bound surface charges from hydration stabilisation which arises because the bound charges encourage the binding of water molecules to the particle surface.

Exclusion stabilisation has only recently been recognised as a factor which can contribute to the stability of lyophobic colloids. It is brought about by the presence of dissolved macromolecules in the dispersion medium. Its origin is essentially the non-availability of certain conformations to the dissolved macromolecules in the region between two particles as those particles approach one another closely. Exclusion stabilisation of a rubber latex is obviously possible only if macromolecular substances are present dissolved in the aqueous phase of the latex. However, even if such substances are present, it is not clear as yet whether or not their contribution to the colloid stability of the latex is significant in relation to the contributions from other factors.

2.2. Destabilisation of Rubber Latices

Although it is essential that rubber latices should be sufficiently colloidally stable to be able to persist almost unchanged at normal temperatures for periods of months, if not years, it is also necessary for many applications that they should be capable of being colloidally destabilised in a controlled manner at certain stages of their processing. For most applications where colloidal destabilisation is necessary, it is essential that this stabilisation should take the form of *gelation*, i.e., the latex becomes transformed from a fluid to a uniform gel. Gelation may be accompanied by *syneresis*, i.e., by spontaneous contraction of the gel with exudation of some of the aqueous phase. Destabilisation may also take two other forms, which are commonly referred to as *coagulation* and *flocculation*, respectively. By the former term is meant the formation of discrete macroscopic lumps of polymer; the latter term is used to denote a process which is in effect micro-coagulation. The factors which determine whether a given destabilisation process results in gelation, coagulation or flocculation in rubber latices are generally not well understood. However, it seems to be a necessary condition for gelation that the influence which is responsible for destabilisation should be more-or-less uniformly present throughout the bulk of the latex.

The influences which lead to the destabilisation of rubber latices can be classified under two broad headings:

(1) those influences which lead to a reduction of the potential-energy barrier which discourages the close approach of two particles

(2) those influences which enable the particles the more readily to overcome the potential energy barrier which exists between them

Roughly speaking, this division corresponds, respectively, to destabilisation by chemical influences on the one hand and to destabilisation by physical influences on the other. However, it would be unwise to press these parallels too closely.

2.3. Destabilisation of Rubber Latices by Chemical Influences

It is now necessary to consider in more detail the effects of those destabilisative influences which operate by essentially chemical mechanisms. These influences, which are known generally as *coagulants* or *coacervants*, can be conveniently classified under three headings:

(1) *direct* or *contact coagulants*, which are substances which bring about an immediate and evident destabilisation more-or-less as soon as they are added to the latex

(2) *heat-sensitising coagulants*, which are substances which have relatively little effect upon the stability of a rubber latex at room temperature, but which cause the latex to destabilise rapidly if the temperature is raised above a certain value

(3) *delayed-action coagulants*, which are substances which have relatively little effect upon the stability of a rubber latex when first added to the latex, but which cause the latex to destabilise rapidly after a certain time has elapsed

In the following discussion of these various classes of chemical coagulant, it is necessary to limit consideration to behaviour in anionic rubber latices which are stabilised primarily by the presence of adsorbed or combined carboxylate groups present at the rubber–aqueous phase interface, and whose aqueous phases are alkaline. This type of rubber latex in fact embraces many of those which are of industrial importance. It will also become evident from the following discussion that in some instances the division of coagulants into direct coagulants, heat-sensitising coagulants and delayed-action coagulants is rather arbitrary.

2.3.1. Direct or Contact Coagulants

These are of five distinct types. The first type comprises *acidic substances*. Their effect is attributed principally to interaction between hydrogen ions derived from the acidic substance and the carboxylate ions which stabilise the latex:

$$RCO_2^- + H^+ \rightarrow RCO_2H$$

The result is the formation of the corresponding unionised carboxylic acid. As a consequence, the electric charge at the particle surface is lost, as also is any hydration which is associated with the presence of that charge. The usual effect of adding acidic substances to rubber latices is to cause the rapid formation of lumps of coagulum. However, small quantities of acids and latent acids can sometimes be added without any immediate apparent change, although a slow gelation may occur subsequently.

The second type of direct coagulant comprises *water-soluble salts of heavy metals*. Two types of reaction may occur when aqueous solutions of such salts are added to a carboxylate-stabilised rubber latex whose aqueous phase is alkaline. Direct interaction with the stabilising carboxylate ions may occur, with the formation of insoluble, unhydrated and probably unionised metal carboxylates:

$$M^{n+} + nRCO_2^- \rightarrow (RCO_2)_n M$$

Alternatively, an insoluble heavy-metal hydroxide may be precipitated in the aqueous phase:

$$M^{n+} + nOH^- \rightarrow M(OH)_n$$

Both these reactions can cause destabilisation of the latex. The first causes the particles to lose both charge and hydration. The second results in desorption of stabilisers from the particles because the stabilisers become adsorbed on to the precipitated hydroxide. The second reaction may also result in some mechanical entanglement or co-precipitation of the polymer particles with the insoluble hydroxide. Water-soluble salts of bivalent metals such as calcium, magnesium and zinc tend to cause rapid coagulation of carboxylate-stabilised rubber latices, the coagulum produced being tough and coherent. Water-soluble salts of trivalent metals such as aluminium and ferric iron tend to cause the latex particles to flocculate rather than coagulate.

The third type of direct coagulant comprises *water-miscible organic liquids*, such as ethyl alcohol and acetone. They tend to cause rapid coagulation of rubber latices, principally because they cause the particles to lose bound water, but also because they reduce the dielectric constant of the aqueous phase, thereby causing some compression of the electrical double layers surrounding the particles.

The fourth type of direct coagulant comprises *rubber-miscible organic liquids*, such as benzene and carbon tetrachloride. These liquids become gradually imbibed into the rubber particles, thereby causing the rubber particles to swell and integrate with each other to form a coherent coagulum. In effect, imbibition of the organic liquid by the rubber particles causes the volume fraction of disperse phase in the latex to increase. It will be noted that in this case the mechanism of destabilisation is physical rather than chemical. It is also questionable whether these organic liquids should really be classified as direct coagulants, because their action is rather slower than that of most of the other direct coagulants. It would perhaps be more logical to class them as delayed-action coagulants, although it is not usual to regard them as such.

The fifth type of direct coagulant comprises the *cationic surfactants*. They can cause rapid destabilisation by becoming adsorbed at the particle surfaces, and thereby reducing both surface charge and probably also surface hydration. Destabilisation by added cationic surfactants is frequently manifest by progressive thickening followed by flocculation or coagulation. However, if a large excess of a cationic surfactant is rapidly added to an anionic rubber latex, then sufficient surface-active cations may become adsorbed to bring about reversal of the charge carried by the particles. The product is then a colloidally-stable rubber latex whose particles carry positive electric charges.

2.3.2. Heat-Sensitising Coagulants

The best-known method of making carboxylate-stabilised rubber latices heat-sensitive is by means of the so-called *zinc ammine system*. This method requires that there should be present in the latex a source of zinc ions which is sparingly soluble in the aqueous phase, free ammonia, and ammonium ions. It is also necessary that the latex should be susceptible to destabilisation by zinc ions. If these conditions are fulfilled, a very slow thickening and gelation takes place at room temperature. This phenomenon is commonly known as *zinc oxide thickening*. Thus the zinc ammine system also displays something of the character of a delayed-action coagulant as well as functioning as a heat-sensitising coagulant, providing an illustration of the rather arbitrary nature of the distinction between these two types of coagulant. The rate of gelation increases sharply as the temperature is raised, gelation becoming very rapid at temperatures in excess of 50°C. This method of heat-sensitising rubber latices has been most widely used with natural rubber latex, partly because it was first developed for use with this latex, and partly because, as will appear subsequently (Section 3.2.3), natural rubber latex as available commercially frequently contains ammonia as a preservative. The basic chemistry underlying heat-sensitisation by the zinc ammine system is not fully understood, but the following reaction scheme, in which the source of zinc ions is taken to be zinc oxide, probably represents approximately what happens:

$$\text{ZnO} \overset{\text{H}_2\text{O}}{\rightleftharpoons} \text{Zn(OH)}_2 \rightleftharpoons \text{Zn}^{2+} + 2\text{OH}^-$$

$$\Big\Updownarrow {\scriptstyle \text{NH}_3}$$

$$\text{Zn}^{2+} + n\text{NH}_3 \overset{\text{heat}}{\longleftarrow} \text{Zn(NH}_3)_n^{2+}$$
$$\searrow {\scriptstyle \text{RCO}_2^-} \qquad 1 \leqslant n \leqslant 4$$

$$(\text{RCO}_2)_2\text{Zn}$$

insoluble zinc soap or
proteinate precipitated
in surface of particles,
with consequent
destabilisation

In the presence of water, the zinc oxide behaves as though it were zinc hydroxide. The zinc ions which are liberated are complexed by the free ammonia to give zinc ammine ions of various compositions. When the temperature of the latex is raised, these zinc ammine ions tend to lose ammonia to form lower zinc ammine ions and, ultimately, hydrated zinc ions. Either the resultant free zinc ions or the lower ammine ions (or possibly both) then interact with the carboxylate ions which stabilise the latex. Insoluble zinc derivatives of these carboxylate ions then form, causing the latex to become destabilised. The function of the ammonium ions, which do not enter explicitly into the above scheme of reactions, is to suppress the concentration of hydroxyl ions, and so, by the solubility product principle, to enhance the initial concentration of zinc ions. Likewise, the heat-sensitivity is reduced or inhibited altogether by the addition of strong alkalis such as potassium hydroxide, and also by the addition of excessive amounts of ammonia. Perhaps rather surprisingly, the degree of heat-sensitisation imparted to natural rubber latex by the zinc ammine system is increased by the

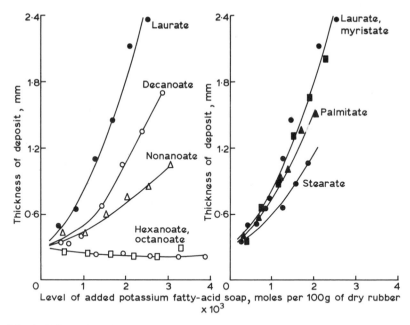

Fig. 1. Effect of added potassium n-alkanoates upon heat-sensitivity of natural rubber latex containing zinc ammine ions.[5] Heat-sensitivity was assessed by determining the thickness of rubber deposited on a heated former immersed in the latex for a given time.

addition of fatty-acid soaps. The enhancement of heat-sensitivity depends not only upon the amount of fatty-acid soap which is added, but also upon the nature of the soap. Results obtained by Blackley and Asiah[5] for the effects of added potassium n-alkanoates are summarised in Fig. 1; rather interestingly, the maximum effect is observed when the alkyl chain of the n-alkanoate contains approximately 11 carbon atoms. More detailed discussions of the mechanism of heat-sensitisation by the zinc ammine system will be found in papers by Kraay and van den Tempel,[6] by McRoberts,[7] and by van den Tempel.[8] A detailed experimental investigation of matters relevant to the heat-sensitisation of rubber latices by the zinc ammine system is described in a thesis by Nithi-Uthai.[9]

It may be noted in passing that the progressive thickening which is observed to occur when ammonia-preserved natural rubber latex to which zinc oxide has been added is allowed to stand at room temperature is believed to have its origin in the same processes as those which are responsible for heat-sensitisation by the zinc ammine system. Those factors which enhance heat-sensitisation also enhance thickening, and those factors which discourage heat-sensitisation also discourage thickening. In particular, it is found that the rate of thickening is enhanced by the addition of fatty-acid soaps.

As alternatives to the zinc ammine system, various other heat-sensitisers are available which depend for their action upon being soluble in cold water but insoluble in hot water. The best-known heat-sensitising additive of this type is poly(vinyl methyl ether) (I).

$$\ldots -\underset{\underset{\underset{\textrm{I}}{CH_3}}{\overset{|}{\underset{|}{O}}}{CH} - CH_2 - \ldots$$

Although soluble in cold water, this polymer precipitates almost completely from aqueous solutions when the temperature is raised to 33–34°C. The precipitation temperature is virtually independent of concentration, polymer molecular weight, pH and ionic strength. The ability to heat-sensitise rubber latices is undoubtedly associated with the reduction in water solubility which occurs when the temperature of the latex is raised. However, the precise nature of the connection is not entirely clear. Various suggestions have been made, of which the following are the more credible:

(1) The poly(vinyl methyl ether) remains dissolved in the aqueous phase of the latex, where it precipitates and mechanically entraps the rubber particles in the precipitate.

(2) The poly(vinyl methyl ether) remains dissolved in the aqueous phase of the latex, where it precipitates and destabilises the latex by adsorbing the latex stabilisers on to the precipitate.

(3) The poly(vinyl methyl ether) is adsorbed on to the latex particles. It functions as a steric/hydration stabiliser at room temperature, but its ability to stabilise is lost as the temperature is raised because the tendency of the vinyl methyl ether segments to mix with the aqueous phase of the latex is greatly reduced.

(4) The poly(vinyl methyl ether) remains dissolved in the aqueous phase of the latex at room temperature, but becomes strongly adsorbed on to the latex particles as the temperature is raised,

thereby promoting inter-particle bridging. It is the inter-particle bridging which causes the latex to destabilise.

Of these theories, (3) and (4) seem to be more consistent with the facts than do (1) and (2). The author tends to favour a theory of type (4) rather than one of type (3). Poly(vinyl methyl ether) is principally used as a heat-sensitiser for natural rubber latex. The heat-sensitivity of natural rubber latex which contains poly(vinyl methyl ether) increases as the pH is reduced, and is also increased if zinc oxide is present. Both these observations are consistent with the view that increased rubber–aqueous phase interface is thereby made available for the adsorption of the polymer. Even in the presence of zinc oxide, the heat-sensitising efficiency of poly(vinyl methyl ether) is low at the normal pH (ca 10·3) of ammonia-preserved natural rubber latex which contains about 0·6% ammonia. However, if the pH of the latex is reduced to a value between 9·6 and 9·0, then a very heat-sensitive latex is obtained if zinc oxide is also present. In the absence of zinc oxide, the pH must be reduced further if appreciable heat-sensitivity is to be obtained. The degree of heat-sensitisation also depends upon the molecular weight of the poly(vinyl methyl ether); the higher is the molecular weight of the polymer, the greater is the heat-sensitivity imparted under given conditions. A great advantage which poly(vinyl methyl ether) has over the zinc ammine system as a heat-sensitiser for natural rubber latex is that, in contrast to the latter, it functions as a colloid stabiliser at room temperature. Further information concerning poly(vinyl methyl ether) as a heat-sensitiser for natural rubber latex is available in a paper by Cockbain[10] and in a recent thesis by Sivagurunathan.[11]

Other substances which function as heat-sensitisers for rubber latices in a similar manner to poly(vinyl methyl ether) include polypropylene glycols (**II**) of low molecular weight (500 to 800), certain water-soluble

$$\ldots -CH-CH_2-O- \ldots$$
$$\quad\quad | $$
$$\quad\quad CH_3$$

II

siloxane polymers, and certain non-ionic surfactants of the poly(ethylene oxide) type. The possibility of using substances of the latter type as heat-sensitisers for natural rubber latex has recently been investigated in some detail by Gorton.[12] Heat-sensitisers of the siloxane type have been developed specifically for use with synthetic rubber latices.

2.3.3. Delayed-Action Coagulants

By far the most important type of delayed-action coagulant for rubber latices is the *alkali-metal salts of hydrofluorosilicic acid* (H_2SiF_6), which are commonly known as *silicofluorides*. They depend for their action upon the slow, reversible hydrolysis of the silicofluoride anion as follows:

$$SiF_6^{2-} + 4H_2O \rightleftharpoons [Si(OH)_4] + 6F^- + 4H^+$$

unstable ortho-
silicic acid

self-condensation with
elimination of water

$$\ldots-O-\underset{\underset{OH}{|}}{\overset{\overset{O}{|}}{Si}}-O-\underset{\underset{OH}{|}}{\overset{\overset{OH}{|}}{Si}}-OH$$

precipitate of gelatinous hydrated silica

The orthosilicic acid, which is probably the immediate product of the hydrolysis, is a very unstable substance which readily condenses with itself to give a gelatinous, colloidal precipitate which is loosely described as 'colloidal silica'.

The precise mechanism whereby the silicofluorides bring about gradual destabilisation of carboxylate-stabilised rubber latices is not entirely clear, but at least five processes are probably involved:

(1) The pH of the aqueous phase of the latex is gradually lowered by the liberation of hydrogen ions.

(2) The latex particles become mechanically entrapped in or co-precipitated with the gelatinous silica which forms.

(3) The latex stabilisers become adsorbed on to the precipitated gelatinous silica.

(4) As the gelatinous silica precipitates, so it becomes adsorbed on to the rubber particles, thereby causing inter-particle bridging.

(5) The lowering of the pH encourages the dissolution of zinc, probably in the form of zinc ammine ions if ammonia is present, or as other zinc complex ions if, as is sometimes the case, other nitrogenous substances are present. Silicofluorides are almost always used in rubber latices in conjunction with a sparingly

soluble zinc compound, such as zinc oxide, in order to ensure the formation of firm, coherent gels. Increased dissolution of zinc as the pH falls is thought to be an important factor in encouraging the formation of such gels.

The delayed-action effect of the alkali-metal silicofluorides is attributable in part to the slow rate at which the silicofluoride anion hydrolyses, but is also attributable in part to the low solubility of these silicofluorides in aqueous media. (Rather unusually, the alkali–metal silicofluorides are less soluble in water than are some of the heavy-metal silicofluorides.)

Of the various alkali–metal silicofluorides, it is the sodium salt which has been most widely used as a delayed-action coagulant for rubber latices. The potassium salt has also been used; it is rather slower in action than is the sodium salt, because it is less soluble in water. If the conditions of destabilisation are correctly adjusted, rubber latices containing these silicofluorides will set to firm, coherent gels over a period of a few minutes. The rate of gelation, and the nature of the gel obtained, depend upon several factors, such as the initial alkalinity and pH, the levels and types of surfactants which are present in the latex, the presence or absence of sparingly soluble zinc compounds, and the temperature. As would be predicted from the effects of temperature upon the solubility of the silicofluorides and upon the rate of hydrolysis of the silicofluoride anion, increase of the temperature causes the rate of gelation to increase. Thus it is that the silicofluorides also display something of the character of heat-sensitising coagulants as well as functioning as delayed-action coagulants. They therefore provide a further illustration of the rather arbitrary nature of the distinction between the two types of coagulant. Further information on the behaviour of silicofluorides as delayed-action coagulants for rubber latices can be found in papers by McKeand[13] and by Madge.[14]

Some interest has also been shown in the use of other analogous fluoro salts, such as sodium borofluoride ($NaBF_4$), sodium titanofluoride (Na_2TiF_6) and sodium zirconofluoride (Na_2ZrF_6).[13,15] As far as is known, the use of these substances has remained of academic interest only. Presumably the titanofluorides and the zirconofluorides function by a similar mechanism to that by which the silicofluorides effect gradual destabilisation. Amongst the various other systems which have been proposed as delayed-action coagulants for rubber latices are combinations of persulphates and reducing agents.[16] These function by the gradual formation of hydrogen ions in the aqueous phase of the latex according to a reaction such as:

$$S_2O_8^{2-} + 2[H] \rightleftharpoons 2HSO_4^- \rightleftharpoons 2H^+ + 2SO_4^{2-}$$

Suitable reducing agents include sodium thiosulphate and hydroxylamine. The latter may be added as hydroxylammonium chloride, in which case an additional equivalent of acid is liberated, derived from the hydrochloric acid which is combined in this salt.

2.4. Destabilisation of Rubber Latices by Physical Influences
Roughly speaking, physical agencies are able to bring about destabilisation of a rubber latex if they increase either the frequency or the violence of the collisions between the particles, or both, or if they reduce the volume of dispersion medium which is available to the particles. Physical influences which bring about the destabilisation of rubber latices can be conveniently classified under three headings:

(1) heating
(2) freezing
(3) mechanical agitation

2.4.1. Destabilisation by Heating
Raising the temperature of a latex increases both the frequency and the violence of the collisions between the particles, and therefore tends to increase the rate of flocculation and coagulation. Of these two factors, the increase in the violence of the collisions is the more important. Increasing the temperature of a latex may also bring about desorption and denaturation of the stabilisers, as well as partial dehydration of the particles. All these factors will tend further to reduce the colloid stability of the latex.

2.4.2. Destabilisation by Freezing
For the same reasons as have been advanced in Section 2.4.1, lowering of the temperature of a rubber latex should enhance its colloid stability. This is in fact generally the case until the temperature is reduced to such an extent that the aqueous phase of the latex begins to freeze. The colloid stability of the latex is then greatly reduced for two distinct reasons:

(1) Because it is ice crystals and not frozen latex which separate initially, the volume of dispersion medium which is available to

the particles becomes progressively reduced and the collision frequency rises accordingly.

(2) As the volume of unfrozen aqueous phase becomes reduced, so rubber particles tend to become entrapped between growing ice crystals and are thereby subjected to high pressures. Under these conditions, agglomeration and coalescence of the particles occurs rapidly.

As when a rubber latex is heated, the actual changes which occur when a rubber latex is cooled and partially frozen will also be determined in part by any chemical and physico-chemical changes involving the latex stabilisers which may occur. The subject of the freeze–thaw stability of polymer latices has been thoroughly reviewed in a thesis by Teoh.[17] This thesis also contains an account of an extensive experimental investigation of the factors which influence the freeze–thaw stability of polymer latices. A summary of this investigation is also available in a paper by Blackley and Teoh.[18] The matter of the ability of a rubber latex to withstand repeated cycles of freezing and thawing without becoming colloidally destabilised is of considerable industrial importance in relation to the storage and transportation of rubber latices during winter months.

2.4.3. Destabilisation by Mechanical Agitation

Mechanical agitation of rubber latices causes the particles to collide more frequently and more violently than in the un-agitated state, and therefore tends to destabilise the latex. This matter is again of great industrial importance because of the need to pump and transport rubber latices. It is also of importance in connection with the manufacture of latex foam rubber. The term 'mechanical stability' is used to denote the ability of a latex to withstand mechanical agitation without becoming colloidally destabilised. The mechanical stability of rubber latices is profoundly affected by certain variables such as the levels and types of surfactants which are present in the latex, the pH of the latex, and the levels and types of any electrolytes which may be present in the latex. Figure 2 summarises the results of Blackley et al.[19] for the effects of added potassium n-alkanoate soaps of various alkyl chain lengths upon the mechanical stability of ammonia-preserved natural rubber latex. At any given level of addition, the optimum enhancement of mechanical stability is observed when the alkyl chain length of the soap is approximately 11. Very large enhancements of mechanical

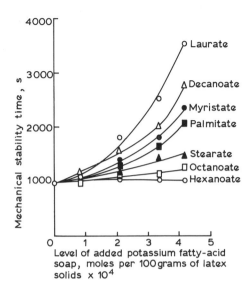

Fig. 2. Effect of added potassium n-alkanoates upon mechanical stability of natural rubber latex.[19]

stability are brought about by small additions of soaps such as potassium laurate; this effect has been attributed to activation of the indigenous soaps as stabilisers rather than to the increase in the soap content of the latex *per se*. The effects of alkalinity and of addition of electrolytes upon the mechanical stability of natural rubber latex are shown in Figs 3[20] and 4,[21] respectively; the drastic effect of heavy metal ions in reducing the mechanical stability of ammonia-preserved natural rubber latex should be noted. Greene and co-workers[22,23] have made the interesting observation that carboxylate anions which are chemically combined at the surface of the latex particles are more efficient in conferring mechanical stability upon synthetic rubber latices than are adsorbed carboxylate anions.

Extensive reviews of the literature pertaining to the mechanical stability of rubber latices will be found in theses by Teoh[17] and by Tan.[20] A paper by Pendle and Gorton[24] reviews the mechanical stability of natural rubber latex specifically. A paper by Roe and Brass[25] contains useful information concerning the mechanical stability of synthetic latices, although it is concerned with polystyrene latices rather than with rubber latices. Theoretical considerations of some of

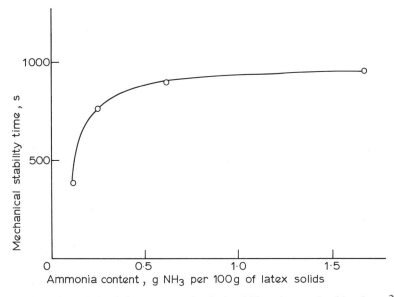

Fig. 3. Effect of alkalinity upon mechanical stability of natural rubber latex.[20]

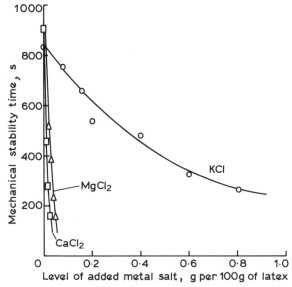

Fig. 4. Effect of added metal salts upon mechanical stability of natural rubber latex.[21]

the factors which affect the mechanical stability of latices have been given by Utrachi.[26]

3. NATURAL RUBBER LATEX

3.1. Introduction

Natural rubber latex is derived from the fluid which is found in certain specialised cells of trees of the *Hevea brasiliensis* species. These cells form a network of interconnecting vessels which surround the trunk of the tree. They are found in the inner cortex region. The latex is obtained from the tree by severing the latex vessels, whereupon the latex exudes from the cut spontaneously. Natural rubber latex as obtained from the tree is a complex biological fluid, whitish in colour, of pH between 6·5 and 7·0, which is of variable composition and viscosity. It contains approximately 33% by weight of rubber hydrocarbon dispersed as particles, together with minor amounts of other particulate phases. The distribution of particle sizes amongst the rubber particles is very wide. The overall total solids content is approximately 36%. In addition to rubber hydrocarbon and water, fresh natural rubber latex contains a variety of other substances such as proteins, lipids, carbohydrates and resinous substances. More detailed information concerning the composition of fresh natural rubber latex is available in a book edited by Bateman.[27]

The rubber hydrocarbon contained in natural rubber latex is high-molecular-weight linear *cis*-1,4-poly(isoprene). Much interest has been shown in the biochemical pathway by which this hydrocarbon is produced in the *H. brasiliensis* tree. The hydrocarbon is certainly not produced by the addition polymerisation of isoprene. It is now well established that it is in fact produced by the condensation polymerisation of isopentenyl pyrophosphate. An extensive review of the biosynthesis of natural rubber has been published by Fournier and Tuong-Chi-Cuong.[28] A very brief review has also been published recently by Archer *et al.*[29]

A matter which is cognate to the biosynthesis of natural rubber concerns the role which natural rubber plays in the life of those plants which produce it. Several suggestions have been made, including:

(1) that it provides a wound-healing mechanism when the plant is damaged
(2) that it is a reserve food supply

(3) that it is a metabolic end-product of a process which leads to the production of other useful end-products for the plant

However, despite much speculation, the true role of natural rubber in the plants which produce it remains obscure.

3.2. Preservation of Natural Rubber Latex

3.2.1. Why Preservation is Necessary

Fresh natural rubber latex coagulates within a few hours of leaving the tree. The actual time required for coagulation to occur depends very much upon the ambient temperature and upon the stability of the latex itself. The result is separation into clots of rubber and a clear serum. At a later stage, putrefaction sets in, with the development of bad odours. It is to prevent both of these processes that preservation is necessary.

A distinction is made between long- and short-term preservation. The former is intended to imply preservation during the period of shipment to, and transportation and storage within, the consumer country. The latter implies merely preservation sufficient to ensure that the field latex remains in a liquid condition for a few hours or a few days before being processed into the various forms of dry rubber. Short-term preservatives are commonly known as *anticoagulants*.

Two general theories have been proposed to account for the phenomenon of the spontaneous coagulation of fresh natural rubber latex. The first supposes that the effect is primarily a consequence of the development of acidity through micro-organisms interacting with various non-rubber constituents. The second attributes the effect to the liberation of fatty-acid anions through the hydrolysis of various of the lipid substances present in the latex. Such anions are then thought to be adsorbed on to the surfaces of the rubber particles, possibly partially displacing adsorbed proteins. They then interact with metal ions, such as those of calcium and magnesium, which are either present in the latex initially or else are gradually released by the action of enzymes. Of these two theories, the second appears to accord rather better with the facts than does the first.

3.2.2. Requirements of the Ideal Preservative

The primary requirement of an ideal preservative for natural rubber latex is clearly that it should preserve the latex efficiently against spontaneous coagulation and putrefaction. At least three distinct ways in which the preservative should operate can be distinguished:

(1) it should destroy micro-organisms, or at least suppress their activity and growth
(2) it should enhance the colloid stability of the latex
(3) it should deactivate trace metals, especially traces of heavy-metal ions, either by sequestration in solution or by precipitation as insoluble salts

As regards enhancement of colloid stability, because of the nature of natural rubber latex, this is most readily achieved if the preservative is alkaline in reaction. This is because the protective proteins in natural rubber latex are on the alkaline side of their isoelectric points when the latex leaves the tree, so that increasing the pH of the latex tends to increase the charge carried by the particles. Deactivation of heavy-metal ions is important for two reasons. In the first place, some of these metal ions are essential for the metabolism of those micro-organisms which cause spontaneous coagulation. Secondly, such ions tend to destabilise the latex.

The ideal preservative will also meet a number of ancillary requirements as well as being effective as a preservative. Thus it should obviously be harmless to both people and rubber, and its use should not give rise to environmental problems. It should not discolour the latex, or films deposited from it. It should not impart an offensive odour. It should not interfere with established processes. It should be cheap and convenient to handle.

3.2.3. Ammonia as a Preservative for Natural Rubber Latex

The best-known preservative for natural rubber latex is ammonia. It is usually added to the latex as a gas from cylinders, rather than as an aqueous solution. 0·2% of ammonia on the whole latex is sufficient for short-term preservation; approximately 0·7% is required for long-term preservation. Ammonia is very effective as a bactericide if the level exceeds 0·35% on the whole latex. It imparts an alkaline reaction to the latex, thereby enhancing the colloid stability of the latex. The effect of ammoniation upon the pH of natural rubber latex is shown. in Fig. 5. Ammonia is able partially to sequester certain heavy-metal ions by ammine formation. Others it precipitates as insoluble hydroxides. It is particularly significant that any soluble magnesium ions present in the

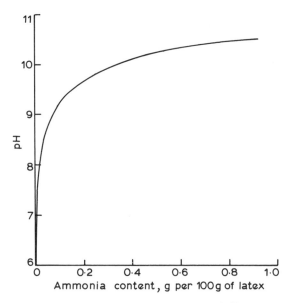

Fig. 5. Effect of ammonia concentration upon pH of 60% natural rubber latex.

latex are precipitated as the very sparingly-soluble magnesium ammonium phosphate, if sufficient phosphate ions are also present:

$$Mg^{2+} + NH_3 + HPO_4^{2-} \rightarrow MgNH_4PO_4$$

In fact, fresh natural rubber latex does usually contain significant levels of soluble magnesium ions, and, if left in this form, these ions can seriously reduce the colloid stability of the latex for the reason given in Section 2.3.1. There seems little doubt that one important reason for the long-standing success of ammonia as a preservative for natural rubber latex is its ability to precipitate magnesium as magnesium ammonium phosphate.

Thus ammonia fulfils the primary criteria for an efficient preservative very well. Unfortunately, it does not meet the ancillary requirements so well. In particular, it is inconvenient to handle, it is not particularly cheap, it has a strong odour, it interferes with certain aspects of latex processing (notably gelation by silicofluorides), it promotes zinc oxide thickening, and its subsequent disposal presents environmental problems. Although many attempts have been made to find alternative substances which will completely replace ammonia as a general-purpose preservative for natural rubber, it has so far proved possible merely to reduce the ammonia level and incorporate secondary preservatives. Such latices are commonly known as 'low-ammonia' latices. Typical combinations which have been used industrially include:

(1) 0·2% sodium pentachlorophenate + 0·2% ammonia
(2) 0·25% boric acid + 0·05% sodium pentachlorophenate + 0·2% ammonia
(3) 0·1% zinc diethyl dithiocarbamate + 0·2% ammonia
(4) less than 0·05% tetramethylthiuram disulphide + not more than 0·02% zinc oxide + 0·2% ammonia

Attempts have recently been made to reduce the ammonia level still further using tetramethylthiuram disulphide and zinc oxide as secondary preservatives. It appears that the ammonia content can be reduced to as low as 0·05% if a small amount of potassium hydroxide is also added to supplement the alkalinity. Further information concerning the older types of low-ammonia natural rubber latex is available in a paper by Philpott;[30] the newer types are described in two papers by Calvert et al.[31,32]

The ammoniation of fresh natural rubber latex brings about important chemical changes, mainly as a consequence of the concomitant increase in the pH of the aqueous phase. These changes include:

(1) hydrolysis of proteins to peptides and amino acids
(2) hydrolysis of phospholipids to fatty-acid anions, glycerol, phosphate anions and amines.
(3) destruction of the minor particulate phases present in fresh natural rubber latex

One consequence of these changes is that, whereas fresh natural rubber latex owes its colloid stability largely to the presence of proteins adsorbed at the surfaces of the rubber particles, ammonia-preserved natural rubber latex is stabilised partly by adsorbed fatty-acid anions and partly by proteinaceous residues. A detailed discussion of the composition of ammonia-preserved natural rubber latex will be found in the book edited by Bateman.[27]

3.2.4. Potassium Hydroxide as a Preservative for Natural Rubber Latex

The only sole preservative which has been widely used for natural rubber latex other than ammonia is potassium hydroxide. This has been used as a preservative for a very high-solids concentrate (*ca* 73% total solids content) which is made by evaporation (see Section 3.3.2). Potassium hydroxide owes its effectiveness as a preservative principally to the high pH which it imparts to the latex.

3.3. Concentration of Natural Rubber Latex

3.3.1. Why Concentration is Necessary
Because its rubber content is only approximately 33%, it is not economical to transport preserved field natural rubber latex over large distances. Furthermore, many of the industrial processes for which natural rubber latex is used are not well adapted to latices having rubber contents as low as *ca* 33%. These are the principal reasons why natural rubber latex is now invariably concentrated to a product which has a rubber content of 60% or above. A further advantage which accrues from concentrating natural rubber latex is that the uniformity of the product is thereby improved. This is partly a consequence of the partial removal of non-rubber constituents which is a feature of several of the processes.

3.3.2. Concentration by Evaporation
This process differs fundamentally from the other three concentration processes described below in that water only is removed from the latex. In particular, there is no removal of non-rubber constituents other than water, or of very small rubber particles. In a typical evaporative process for the concentration of natural rubber latex, the latex is heated to 90°C and introduced through suitable nozzles into an evaporation chamber in which the air pressure has been reduced to one-tenth atmospheric. The boiling point of water at such a pressure is *ca* 46°C. At the moment of entering the evaporator, a rapid fall in temperature of *ca* 44°C occurs, and at the same time the water content of the latex is reduced by about 7%. The operation is repeated until the water content has been reduced to the desired level. The best-known natural rubber latex concentrate produced by evaporation has a solids content between 72 and 75%. It is preserved by potassium hydroxide and is stabilised during production by the addition of a small quantity of a potassium soap. An ammonia-preserved concentrate of approximately 62% total solids content has also been produced by evaporation.

3.3.3. Concentration by Creaming

Natural rubber has a lower specific gravity than does the aqueous phase of natural rubber latex. The particles in natural rubber latex therefore tend to rise to the surface if the latex is left undisturbed. This process, which is known as *creaming*, has been used as the basis of a method of concentrating natural rubber latex. In order for the process to be practicable, it is necessary to add substances known as *creaming agents* which accelerate the process. These are water-soluble hydro-colloids such as ammonium alginate. These substances appear to function by causing some form of reversible and temporary agglomeration of the latex particles, such as by inter-particle bridging. The agglomerates cream more rapidly than do the primary latex particles. Concentration by creaming is accompanied by some loss of non-rubber constituents and small rubber particles.

3.3.4. Concentration by Centrifuging

The most important of the methods currently being used for the concentration of natural rubber latex is centrifuging. This is essentially a creaming process in which the rate of movement of the rubber particles relative to the aqueous phase is accelerated by replacing gravity by a centrifugal field. In the centrifuges used industrially for concentrating natural rubber latex, the latex is divided into a number of thin concentric shells by a series of conical bowls. The effect of this is to reduce greatly the distance which a rubber particle has to move in order to enter the region in which the concentrate is accumulating. The mode of operation of a typical latex centrifuge should be evident from the diagrammatic section through the bowl of a de Laval centrifuge shown in Fig. 6. As in the case of creaming, centrifuging removes some of the non-rubber constituents and some of the small rubber particles. Natural rubber latex can be subjected to multiple centrifuging, in which case these effects are even more evident. Some typical properties of once- and multiply-centrifuged natural rubber latex are summarised in Table 1.[20]

3.3.5. Concentration by Electrodecantation

In this method of concentrating natural rubber latex, creaming is accelerated by causing the rubber particles to accumulate at the surface of a semi-permeable membrane by establishing an electric field within the latex. Although this process was used industrially at one time, it is

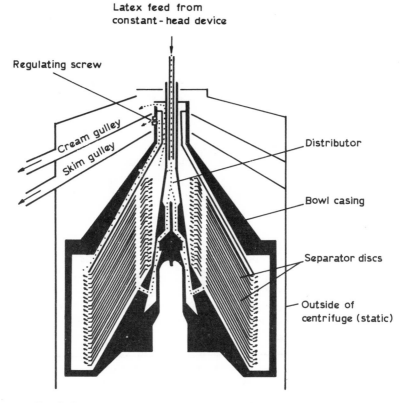

Fig. 6. Schematic cross-section of bowl of de Laval latex centrifuge.

now of academic interest only. Further information concerning this process is available in papers by Murphy[33] and by Stevens.[34]

4. SYNTHETIC RUBBER LATICES

4.1. Introduction

Synthetic rubber latices are produced by the emulsion polymerisation of appropriate monomers. Several extensive reviews of emulsion polymerisation are now available (e.g., Refs 35–41). An English translation of a Russian book edited by Eliseeva has recently become available;[42] in addition to giving a useful résumé of emulsion polymer-

TABLE 1

Properties of Once- and Multiply-Centrifuged Natural Rubber Latices[20]

Property	Number of centrifugings			
	One (LAZN type latex)	Two	Three	Four
Total solids content (%)	61·8	60·8	61·3	60·3
Dry rubber content (%)	60·4	60·1	60·7	59·9
Alkalinity (g ammonia per 100 g aqueous phase)	0·21	0·19	0·19	0·19
KOH number[a]	0·61	0·24	0·15	0·13
VFA number[a]	0·03	0·02	—	—
HFA number[a]	0·02	0·160	0·131	0·110
Mechanical stability time (s)	940	2 160	682	437
Protein content of total solids (%)[b]	1·72	1·04	0·81	0·74
Protein content of dry rubber (%)[b]	1·32	0·87	0·75	0·66
pH	9·90	10·32	10·51	10·59
Electrical conductance at 55% total solids content (mho × 10³)	2·60	1·23	0·75	0·72

[a] The KOH (potassium hydroxide) number, VFA (volatile fatty acid) number and HFA (higher fatty acid) number are, respectively, measures of:

(1) the total level of ammonium salts in the latex
(2) the level of anions of volatile fatty acids (formic, acetic and propionic acids) in the latex
(3) the level of anions of higher fatty acids in the latex

In each case, the levels are expressed as their potassium hydroxide equivalents, the units being the number of grams of potassium hydroxide equivalent to the acid anions which are present in that quantity of latex which contains 100 g of total solids.

[b] The protein contents were calculated as 6·25 times the nitrogen content as determined by the semi-micro Kjeldahl method.

isation as understood by Russian contributors to the subject, it also contains chapters devoted to certain classes of synthetic latex.

Lack of space precludes any detailed discussion here of the emulsion polymerisation reaction as applied to the production of synthetic rubber latices. It will merely be observed that, in certain cases, the temperature at which the polymerisation reaction is carried out is of considerable importance. Especially is this so for the latices of rubbers which contain substantial proportions of combined butadiene. The effect of polymerising at 'low' temperatures (*ca* 5°C) as compared with 'high' temperatures (*ca* 50°C) is to produce polymer molecules which are less branched, and also polymer molecules in which the ratio of *trans*-1,4 enchained butadiene units to *cis*-1,4 enchained butadiene units is increased. These changes have important implications for the properties of the rubbers which are contained in the latices produced.

4.2. Agglomeration and Concentration of Synthetic Rubber Latices
The product of the typical emulsion polymerisation reaction used for the production of synthetic rubber latices contains particles of average diameter between about 50 and 150 nm. The distribution of particle sizes about the average is very narrow; the immediate product of the typical emulsion polymerisation reaction should be contrasted with natural rubber latex (both field and concentrated) in this important respect. The total solids content of the latex varies over the range 35 to 50%, depending upon the recipe which was used and the extent of conversion of the monomers to polymer. Although some interest has been shown in the possibility of concentrating synthetic rubber latices by creaming, the usual method adopted is evaporation. Synthetic rubber latices are not generally amenable to concentration by centrifuging because the particles are too small.

In some cases, the particle size and size distribution are such that it is not possible to concentrate synthetic rubber latices to the desired concentration without them becoming too viscous and pasty for subsequent use. Particularly is this the case for general-purpose styrene–butadiene rubber latices which are intended to compete with natural rubber latex in the manufacture of latex foam rubber. For these latices, it is desirable to concentrate to total solids contents well in excess of 60%, and this is not possible by merely removing water from the latex. Industrial practice is to agglomerate such latices partially and in a controlled manner, so as to increase the average particle size and broaden the particle-size distribution. Usually the agglomeration is effected as a separate step prior to concentration, but, in at least one process, agglomeration and concentration are effected simultaneously. But whether the two processes take place consecutively or concurrently, the idea is that, by coarsening the particle size distribution, the particles will pack together in the concentrated latex more efficiently, thereby giving a product of lower viscosity than would be the case had

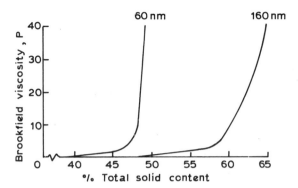

Fig. 7. Effect of particle size upon viscosity–solids relationship for non-carboxylated styrene–butadiene rubber latex.[43]

the latex not been partially agglomerated. The advantageous effect of increasing the particle size upon the viscosity–solids relationship for a styrene–butadiene rubber latex is illustrated in Fig. 7.[43] An important variant which is in principle possible with all agglomeration processes is the process known as *co-agglomeration*. In this process, a blend of a synthetic rubber latex and a synthetic latex of a glassy polymer is co-agglomerated to give an agglomerated latex whose particles contain particles of the glassy polymer embedded in a rubber matrix. The objective is to produce a latex which will deposit a rubber which is reinforced with the particles of glassy polymer.

The extensive patent literature which exists concerning methods of controllably agglomerating synthetic latices bears witness to the industrial importance of the process. The various processes which have been proposed can be classified under the following headings:

(1) agglomeration by solvent addition
(2) agglomeration by partial soap neutralisation
(3) agglomeration by electrolyte addition
(4) agglomeration by freezing
(5) agglomeration by mechanical means
(6) agglomeration by hydrophilic colloids

These processes exploit various of the methods which have been described in Sections 2.3 and 2.4 for colloidally destabilising rubber latices. In practice, the problem is not that of effecting agglomeration, but of effecting it in a controlled and self-terminating manner so that the agglomeration proceeds to a very limited extent only. Of the types of process listed above, it is agglomeration by freezing and agglomeration by mechanical means which have been most widely used. Agglomeration by mechanical means depends upon the fact that a controlled degree of agglomeration can be achieved by passing the latex through a homogeniser. In contrast to most of the other types of method, agglomeration is not caused by any reduction in the colloid stability of the latex, but rather by the pressures and shear stresses which accompany the cavitation which is induced as the latex is forced under relatively low pressure through a homogeniser orifice. By a suitable adjustment of the process variables, it is possible to prepare latices which have a very favourable viscosity–solids relationship. A typical process for agglomerating and concentrating synthetic latices by this method is illustrated diagrammatically in Fig. 8.[44] More information concerning the process can be found in a paper by Jones.[44]

Concerning the sixth type of agglomeration method listed above, it has been found that, under suitable conditions, certain hydrophilic colloids are able to induce partial agglomeration of the particles in synthetic rubber latices. The hydrophilic colloid which has been most widely used for this purpose is poly(vinyl methyl ether). Unlike the other types of agglomeration process, the agglomeration in this case takes place concurrently with concentration. The poly(vinyl methyl ether) can be made more effective by a treatment which is thought to alter its molecular complexity, in particular the extent to which it is crosslinked. This interesting process is described more fully in a paper

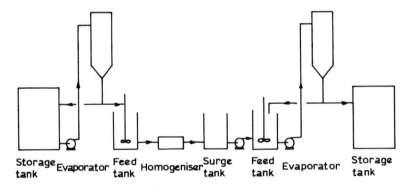

Storage tank   Evaporator   Feed tank   Homogeniser   Surge tank   Feed tank   Evaporator   Storage tank

Fig. 8. Diagrammatic illustration of process for mechanical agglomeration and concentration of non-carboxylated styrene–butadiene rubber latex.[44]

by Howland *et al.*[45] The freeze-agglomeration process is described in a paper by Talalay.[46]

4.3. Individual Types of Synthetic Rubber Latex

For many years now, most of the synthetic rubber latices which have been available commercially fall into four broad groups as follows:

(1) those in which the contained polymer is essentially a copolymer of styrene and butadiene
(2) those in which the contained polymer is essentially a copolymer of acrylonitrile and butadiene
(3) those in which the contained polymer is essentially a homopolymer of chloroprene
(4) those in which the contained polymer comprises essentially units derived from various acrylic monomers

Within each of these four groups, the possibility exists that the polymer may contain minor amounts of other copolymerised monomers which cause the polymer molecules to contain small concentrations of pendant functional groups. The best-known latices of this type are the so-called 'carboxylated' latices, whose polymers are such that they contain minor amounts (say, 1–5% by weight) of copolymerised carboxylic-acid-bearing monomer units. Typically, such units are derived by copolymerising unsaturated acids, such as acrylic acid, methacrylic acid, itaconic acid, and fumaric acid, with the main monomers. Carboxylated synthetic rubber latices have become very important industrially; especially is this the case with those of the styrene–butadiene type.

The industrial advantages which accrue from the carboxylation of synthetic rubber latices fall under two broad headings:

(1) advantages which pertain to the latex
(2) advantages which accrue from modification of the bulk properties of the polymer which is contained in the latex

Under the first of these headings is included advantages such as enhanced mechanical and chemical stability, enhanced tolerance to mineral fillers, and the possibility of preparing polymer latices which can be redispersed after they have been dried down. This last possibility is potentially of great industrial importance; further information is available in a recent paper by Greene *et al.*[47] Under the second of the above headings is included advantages such as susceptibility to reaction with additional reagents, improved adhesion to polar substrates such as textile fibres and metals, enhanced polymer tensile strength, and, to a lesser extent, improved resistance to swelling in hydrocarbon oils. Amongst the additional reagents which are available for reaction with carboxylated polymers are metal oxides, polyamines, urea– and melamine–formaldehyde condensates, and epoxides. Considerable interest has been shown in the use of reactions with such reagents for crosslinking carboxylated rubbers derived from latices. The production of carboxylated latices by emulsion polymerisation is reviewed in a recent paper by the present author;[48] this paper also gives some information concerning the distribution of the carboxylic acid groups in these latices.

Non-carboxylated high-solids styrene–butadiene rubber latices are produced in large quantities as competitors for ammonia-preserved natural rubber latex concentrates in certain applications, notably in the production of latex foam rubber. The weight ratio of styrene to butadiene in these latices is approximately 25/75. The stabiliser is of the fatty-acid soap type, being typically potassium oleate. Carboxylated styrene–butadiene rubber latices of rather lower solids content are widely produced for applications such as paper coating and textile treatments. The styrene/butadiene weight ratio is higher than in the non-carboxylated type, being typically approximately 50/50 or 60/40. The level of carboxylation of the polymer is typically approximately 2%, but the combination of unsaturated acids used to effect carboxylation varies widely. The stabilisers are of the sulphate or sulphonate type; sodium dodecylbenzene sulphonate is a typical example.

Acrylonitrile–butadiene rubber latices, both non-carboxylated and carboxylated, are used in dipping applications where it is desired to manufacture products which show reduced tendency to swell in hydrocarbon oils. Polychloroprene latices have also been used for this purpose, and for the production of dipped products which show enhanced chemical resistance. Carboxylated acrylonitrile–butadiene rubber latices and various acrylic latices are used as binders for bonded fibre fabrics where resistance to oils and fats in the final product is important. Synthetic rubber latices of all types are used for the manufacture of adhesives.

Polychloroprene latices are unusual in that the polymer is susceptible to hydrolysis by the alkaline aqueous medium in which it is dispersed. The hydrolysis reaction involves the minority (*ca* 1·6%) of the chlorine atoms which became enchained in the polymer as allylic chlorine atoms through 1,2 and 3,4 addition during the polymerisation of the chloroprene. The majority of chlorine atoms became enchained as vinylic chlorine atoms through 1,4 addition; these chlorine atoms are resistant to hydrolysis. The result of the hydrolysis of the allylic chlorine atoms is to cause the chlorine atoms to be replaced by hydroxyl groups. There is therefore no main-chain scission of the poly(chloroprene), although its properties are somewhat modified. The pH of the latex gradually falls as hydrolysis proceeds, as does its colloid stability, unless further alkali is added to replace the hydroxyl ions which are lost as a consequence of the hydrolysis.

5. ARTIFICIAL RUBBER LATICES

Many of the large-tonnage and speciality synthetic rubbers are not manufactured by emulsion polymerisation, and, indeed, it is not at present possible to manufacture most of them by this method. The large-tonnage synthetic rubbers of this type include the stereoregular polybutadienes and polyisoprenes, the ethylene–propylene rubbers, and the isobutylene–isoprene (butyl) rubbers. The speciality synthetic rubbers of this type include the chlorosulphonated polyethylenes, the poly(urethane) rubbers, the epichlorohydrin rubbers and the silicone rubbers. Therefore, if the latices of most of these rubbers are required, it is necessary to produce the latices by dispersing the bulk rubber in an aqueous medium. As yet, no really satisfactory way of doing this economically has been found, but potentially there is some industrial interest in this type of latex, especially in those which contain synthetic rubbers of the speciality type, because it would then become possible to manufacture latex products which have the special properties conferred by those rubbers. At the time of writing, as far as is known, industrial interest is confined to latices of isobutylene–isoprene rubbers.

There are three fundamentally distinct processes which have been developed for the preparation of artificial latices:

(1) the *solution-emulsification* technique
(2) the *phase-inversion* technique
(3) the *polymer-neutralisation* technique

In the first of these methods, the rubber is dissolved in, or is swollen by, a volatile solvent. The rubber solution, or swollen gel, is then emulsified in water, and the solvent removed by, say, steam distillation. There is then left behind a dispersion of the polymer in an aqueous medium. In the second method, the rubber is first compounded with a suitable proportion of a long-chain fatty acid such as oleic acid. When the fatty acid is thoroughly dispersed, a dilute aqueous solution of an alkali is slowly worked into the mixture. The initial product is a dispersion of water in rubber, but as more aqueous alkali becomes incorporated, so an inversion of phase occurs, and a rubber-in-water dispersion is produced. In the third method, emulsification is effected by neutralising in the presence of water a rubber whose molecules contain acidic or basic functional groups.

Of these three general methods, most interest has been shown in solution emulsification. However, a major problem which militates against the production of a good latex by this method is the high ratio of colloid stabilisers to rubber which has to be used. The need for this high ratio is a consequence of two factors:

(1) it is a solution of the rubber, and not the rubber itself, which is being dispersed
(2) it is necessary to provide the latex with adequate mechanical and thermal stability to withstand the turbulence and heat of the solvent-stripping process

Two relatively recent processes which exploit the solution-emulsification procedure are the following:

(1) the Exxon process,[49] in which emulsification is effected using a combination of a fatty-acid soap and the salt of a carboxylated α-olefin polymer or copolymer
(2) the Vanderhoff–El-Aasser–Ugelstad process,[50] in which emulsification is effected using a combination of a conventional surfactant and a hydrocarbon, alcohol, amine, etc. of low water solubility and volatility

As an example of the Exxon process may be cited the production of a latex of an isobutylene–isoprene rubber using as emulsifiers a combination of potassium oleate and potassium polyisobutenyl succinate. A typical combination used in the Vanderhoff–El-Aasser–Ugelstad process is sodium dodecyl sulphate and hexadecyl alcohol.

Considerable interest was shown some years ago in a commercially-available artificial latex which was made from synthetic cis-1,4-poly-(isoprene) rubber. This latex was of particular interest because of the similarity between its contained polymer and that which is contained in natural rubber latex. A comparative study of the processing and properties of natural rubber latex and artificial poly(isoprene) latex has been published by Gorton.[51] Artificial poly(isoprene) latex is now largely a curiosity, although it may still be possible to obtain this latex in small quantities commercially.

6. OUTLINE OF APPLICATIONS OF RUBBER LATICES

This chapter will conclude with a very brief review of the major applications of rubber latices. It is convenient to commence this review by noting that it is usually necessary to compound rubber latices before they are used, and that one important facet of compounding is the choice of vulcanising ingredients, i.e., of reactants which are able to effect crosslinking of the rubber molecules when suitable reaction conditions are established. For natural rubber latex, and for synthetic latices which contain copolymerised butadiene units, the primary vulcanising ingredient is usually sulphur. In order that the vulcanisation reaction may proceed at an economically fast rate when dry deposits from the latex are heated at temperatures within the range 100–120°C, it is necessary to include an ultra-accelerator, of which zinc diethyldithiocarbamate is the best-known example. It may also be desirable to include an insoluble zinc compound, such as zinc oxide or zinc carbonate, to activate the vulcanisation accelerator. Other compounding ingredients include fillers, antioxidants, plasticisers, pigments, thickeners and colloid stabilisers. Water-soluble compounding ingredients are usually added to latices as aqueous solutions. Water-insoluble solids are usually added as aqueous dispersions, although for some applications fillers are added dry to suitably stabilised latices. Water-insoluble liquids are usually added as emulsions, but, again, in some cases it is possible to add the liquid to the latex without prior emulsification. It is commonly supposed that water-insoluble solid compounding ingredients—especially vulcanising ingredients—should be dispersed to such an extent that their particle sizes are of a similar order of magnitude to that of the latices to which they are to be added. However, Gorton and Pendle[52–54] have recently produced extensive experimental evidence which compels questioning of this assumption.

In the case of natural rubber latex, the possibility exists of prevulcanising the latex, i.e., of vulcanising the particles in the latex state rather than having to wait until deposits have been dried down from the latex. Prevulcanisation is effected by heating the latex with sulphur and an accelerator. Again the vulcanising ingredients are added to the latex as dispersions, unless they are soluble in water, in which case they are added as solutions. In principle, it is possible to prevulcanise synthetic latices of the styrene–butadiene rubber and acrylonitrile–butadiene rubber types, but in practice the reaction occurs rather slowly, probably because of the absence of non-rubber constituents of the types which are present in natural rubber latex.

As regards the latex applications themselves, we begin with *latex foam rubber.* This is produced by foaming a suitably-compounded latex by mechanical or chemical means, and then causing the latex phase to set to a rubbery continuum. In the older processes, which were widely used for the production of latex foam rubber for cushions, pillows and mattresses, setting was achieved by chemical gelation. The principal methods for effecting the gelation of foamed latices are:

(1) the Dunlop process, which uses alkali–metal silicofluorides as delayed-action coagulants

(2) the Talalay process, in which a mechanically foamed latex is expanded by application of a vacuum, is then fixed by freezing, and is then gelled by allowing an acidic gas (usually carbon dioxide) to diffuse through the foam

(3) the Dow process, in which gelation is effected by reaction between a 'reactive' latex and a 'co-reactive' water-soluble polymer

(4) a process in which the foamed latex is made heat-sensitive by introducing the components of the zinc ammine system, and the foam is then gelled by heating

(5) a process in which the foamed latex is made heat-sensitive by introducing poly(vinyl methyl ether), and the foam is then gelled by heating

The older methods are described in detail in a book by Madge.[55] The Dow process is described in various publications (e.g., Refs 56 and 57); a typical combination of a 'reactive' latex and a 'co-reactive' water-soluble polymer for this latter process would be a carboxylated styrene–butadiene rubber latex and a water-soluble melamine–formaldehyde resin.

Because of adverse competition from polyurethane foam in applications such as cushions, pillows and mattresses, the emphasis in latex foam rubber manufacture has shifted over the last two decades from these products to foam rubber for carpet backing and carpet underlay. This shift from articles of thick cross-section to products of thin cross-section has had important implications for the production technology of latex foam rubber. It has proved practicable to revive an old idea, namely that foamed rubber latex can be set by merely drying it if sufficient foam stabilisers are present to prevent collapse of the foam. Processes of this type are known as *no-gel* latex foam rubber processes. They are clearly unsuited to the production of solid rubber foams of thick cross-section. A detailed description of this type of process will be found in the patent literature (e.g., Ref. 58). One advantage of no-gel processes over gel processes for latex foam rubber manufacture is that a much wider range of foam promoters and foam stabilisers can be used in the case of the no-gel processes; this is because no constraints are placed upon the choice of these components by any requirement for compatibility with a gelation system.

We next consider briefly rubber articles made by *latex dipping*. These include houseware gloves, industrial protective gloves, surgeons gloves, prophylactic sheaths, and toy balloons. The principle of the process is to dip a former of appropriate shape into the latex, remove it, dry and vulcanise the latex deposit on the former, and then remove the deposit from the former. There are four ways in which the latex deposit can be formed upon the former:

(1) by simple or 'straight' dipping without the use of any aid to deposition
(2) by using a coagulant to aid deposition
(3) by using a hot former and a heat-sensitive latex
(4) by electrodeposition, in which the former, which must be metallic, is connected to the positive side of a source of electrical potential and an electrode connected to the negative side of the same source is also immersed in the latex

Most latex dipping is carried out using a coagulant to aid deposition. The coagulant is typically a solution of one or more calcium salts in water or in a mixture of water and industrial ethanol. In a typical process, the former is first immersed in the coagulant solution, is then removed, dried, and then immersed in the latex. Deposition occurs by way of destabilisation by the calcium ions of the coagulant film (see Section 2.3.1). After the former has 'dwelled' in the latex for sufficient time to allow a deposit of the required thickness to be built up, it is removed, and the deposit dried and vulcanised. Although some latex dipping is carried out using synthetic rubber latices of the acrylonitrile–butadiene copolymer and polychloroprene types, most latex dipping is carried out using natural rubber latex. In this case, it is common practice to prevulcanise the latex partially before dipping, in order to reduce the extent of heating which is required to effect full vulcanisation after dipping.

Other applications for rubber latices include binders for non-woven carpets and bonded fibre fabrics, paper treatments, of which paper coating is the most important, water-based adhesives, and the manufacture of rubber thread by extruding latex through an orifice into a coagulant bath. There are many other applications of a less important nature which are too numerous to mention here, but one further major application which must be mentioned is that of the dipping of cords which are to be used in the manufacture of pneumatic tyres. Most tyre-cord dips in use today are based upon blends of a styrene–vinylpyridine–butadiene terpolymer latex and a resorcinol–formaldehyde resin. A styrene–butadiene copolymer latex may also be added to the blend.

REFERENCES

1. Blackley, D. C., *High polymer latices: their science and technology*, (two volumes), Applied Science Publishers Ltd, London (1966).
2. Calvert, K. O. (ed.), *Polymer latices and their applications*, Applied Science Publishers Ltd, London (1982).
3. Blow, C. M., *Proceedings of the Rubber Technology Conference*, London (1938), p. 186.
4. Blow, C. M., *Journal of Society of Chemical Industry*, **57**, 116 (1938).
5. Blackley, D. C. and Asiah, bt. A., *Plastics and Rubber: Materials and Applications*, **4**, 103 (1979).
6. Kray, G. M. and van den Tempel, M., *Transactions of Institution of the Rubber Industry*, **28**, 144 (1952).
7. McRoberts, T. S., *Proceedings of the Second Rubber Technology Conference*, London (1954), p. 38.
8. van den Tempel, M., *Transactions of Institution of the Rubber Industry*, **31**, 33 (1955).
9. Nithi-Uthai, B., PhD Thesis, Council for National Academic Awards (1978).
10. Cockbain, E. G., *Transactions of Institution of the Rubber Industry*, **32**, 97 (1956).
11. Sivagurunathan, L., PhD Thesis, Council for National Academic Awards (1982).

12. Gorton, A. D. T. and Pendle, T. D., *NR Technology*, **11**(1), 1 (1980).
13. McKeand, D. J., *Industrial and Engineering Chemistry*, **43**, 415 (1951).
14. Madge, E. W., *Transactions of Institution of the Rubber Industry*, **28**, 207 (1952).
15. McFadden, G. H. and Bennett, B., *Rubber Age, New York*, **70**, 748 (1952).
16. Twiss, D. F. and Amphlett, P. H., *Journal of Society of Chemical Industry*, **59**, 202 (1940).
17. Teoh, S. C., PhD Thesis, Council for National Academic Awards (1975).
18. Blackley, D. C. and Teoh, S. C., Paper no. 9, *Preprints of PRI International Polymer Latex Conference*, London (1978).
19. Blackley, D. C., Nor Aisah, bt. A. A. and Twaits, R., *Plastics and Rubber: Materials and Applications*, **4**, 77 (1979).
20. Tan, K. Y., PhD Thesis, Council for National Academic Awards (1982).
21. Loha, S., M.Phil. Thesis, Council for National Academic Awards (1977).
22. Greene, B. W., Sheetz, D. P. and Filer, T. D., *Journal of Colloid and Interface Science*, **32**, 90 (1970).
23. Greene, B. W. and Sheetz, D. P., *Journal of Colloid and Interface Science*, **32**, 96 (1970).
24. Pendle, T. and Gorton, A. D. T., *Rubber Chemistry and Technology*, **51**, 986 (1978).
25. Roe, C. P. and Brass, P. D., *Journal of Colloid Science*, **10**, 194 (1955).
26. Utrachi, L. A., *Journal of Colloid and Interface Science*, **42**, 85 (1973).
27. Bateman, L. (ed.), *The chemistry and physics of rubber-like substances*, MacLaren & Sons Ltd, London (1963), especially Chapter 3.
28. Fournier, P. and Tuong-Chi-Cuong, *Rubber Chemistry and Technology*, **34**, 1229 (1961).
29. Archer, B. L., Audley, B. G. and Bealing, F. J., *Plastics and Rubber International*, **7**(3), 109 (1982).
30. Philpott, M. W., *Rubber Developments*, **11**(2), 47 (1958).
31. Calvert, K. O., Sundaram, P. and Tan, T. Y., Paper no. 8, *Preprints of PRI International Polymer Latex Conference*, London (1978).
32. Calvert, K. O., Tan, T. Y., Prichards, R. J. and Sundaram, P., Paper no. 11, *Preprints of PRI Emulsion Polymers Conference*, London (1982).
33. Murphy, E. A., *Transactions of Institution of the Rubber Industry*, **18**, 173 (1942).
34. Stevens, H. P., *Rubber Developments*, **1**(2), 1 (1948).
35. Duck, E. W., In *Encyclopedia of polymer science and technology*, Vol. 5, John Wiley & Sons, New York (1966), p. 801.
36. Blackley, D. C., *Emulsion polymerisation: theory and practice*, Applied Science Publishers Ltd, London (1975).
37. Ugelstad, J. and Hansen, F. K., *Rubber Chemistry and Technology*, **49**, 536 (1976); **50**, 639 (1977).
38. Gardon, J. L., In *Encyclopedia of polymer science and technology*, Supplement Vol. 1, John Wiley & Sons, New York (1976), p. 238.
39. Gardon, J. L., In *Applied polymer science*, J. K. Craven and R. W. Tess (eds), Organic Coatings and Plastic Chemistry Division, American Chemical Society, Washington, DC (1975), p. 138.

40. Dunn, A. S., In *Developments in polymerisation—2*, R. N. Haward (ed.), Applied Science Publishers Ltd, London (1979), p. 45.
41. Piirma, I. (ed.), *Emulsion polymerization*, Academic Press, New York (1982).
42. Eliseeva, V. I. (ed.), *Emulsion polymerization and its applications in industry*, translated from Russian by S. J. Teague, Consultants Bureau, New York (1981).
43. Dunfield, T. E., Watson, W. H. and White, F. C., *Rubber and Plastics Age*, **40**, 1057 (1959).
44. Jones, B. D., *Proceedings of the Fourth Rubber Technology Conference*, London (1962), p. 443.
45. Howland, L. H., Aleksa, E. J., Brown, R. W. and Borg, E. L., *Rubber and Plastics Age*, **42**, 868 (1961).
46. Talalay, L., *Proceedings of the Fourth Rubber Technology Conference*, London (1962), p. 485.
47. Greene, B. W., Nelson, A. R. and Keskey, W. H., *Journal of Physical Chemistry*, **84**, 1615 (1980).
48. Blackley, D. C., *Proceedings of NATO Advanced Study Institute on polymer colloids*, Bristol (1982).
49. Beerbower, A., Burton, G. W. and Malloy, P. L., Exxon Research and Engineering Company, British Patent No. 1,497,757 (1978).
50. Vanderhoff, J. W., El-Aasser, M. S. and Ugelstad, J., US Patent No. 4,177,177 (1979).
51. Gorton, A. D. T., *Rubber Chemistry and Technology*, **43**, 1255 (1970).
52. Gorton, A. D. T. and Pendle, T. D., *NR Technology*, **12**(1), 1 (1981).
53. Gorton, A. D. T. and Pendle, T. D., *NR Technology*, **12**(2), 21 (1981).
54. Gorton, A. D. T. and Pendle, T. D., Paper no. 14, *Preprints of PRI Emulsion Polymers Conference*, London (1982).
55. Madge, E. W., *Latex foam rubber*, MacLaren & Sons Ltd, London (1962).
56. Zimmerman, R. L., Hibbard, B. B. and Bailey, H. R., *Rubber Age, New York*, **98**(5), 68 (1966).
57. Dow Chemical Company, British Patent No. 1,023,202 (1966).
58. Crown Rubber Company, British Patent No. 1,105,538 (1968).

Chapter 8

Colloidal Aspects of Poly(vinyl chloride) Production Processes

D. G. RANCE

*Petrochemicals and Plastics Division, ICI PLC,
Wilton, Cleveland, UK*

1. INTRODUCTION

From the beginning of poly(vinyl chloride) (PVC) production over 40 years ago the polymer has found markets in a large number of industries ranging from building and electrical to furnishing, packaging and clothing. These markets have grown rapidly to make PVC one of today's highest tonnage polymers produced on a world-wide scale. By 1980, the annual world consumption of PVC had reached $1 \cdot 2 \times 10^7$ tonnes per annum. PVC is prepared by four principal polymerisation processes: bulk or mass, suspension, microsuspension and emulsion. In addition, PVC may also be prepared by solution or dispersion polymerisation, but these remain low-tonnage specialised products.

The complexity of PVC manufacturing and processing technology has resulted in the recent publication of excellent volumes entirely devoted to the subject by Burgess[1] and by Butters.[2] This chapter serves to highlight the importance of different colloidal phenomena which occur during the polymerisation processes. The extent to which the processing characteristics of the different types of polymer depend on the careful control of colloidal interactions during polymerisation will be discussed.

Emulsion and microsuspension processes will be considered together in that they both generate aqueous latices. The latices are usually spray-dried before processing. However, the particle size and size distribution of the latex particles play an important role in subsequent processing, particularly when the dried powder is dispersed in a plasticiser to form a so-called plastisol or paste. The types of particle size distribution produced commercially are reviewed, and methods for producing stable latices to high solids content are discussed. This section also considers recent developments in microsuspension polymerisation for producing both monodisperse and polydisperse polymer latices. While spray-drying is commonly used to dry the latex, the energy costs involved in this process are high. Flocculation of the latex, followed by dewatering and air-drying, may provide a cheaper method of polymer isolation.

Suspension and mass polymerisation techniques produce polymer grains with a particle diameter in the range 50–250 μm. In suspension polymerisation, monomer droplets are suspended in an aqueous phase which acts as a heat-transfer medium. The size and size distribution of the droplets are determined by the agitation conditions in the reactor and by the nature and concentration of a macromolecular additive. The type of macromolecule used and its adsorption behaviour at the monomer/water interface have important effects on grain properties. In mass polymerisation the aqueous phase is absent. However, these two processes have in common the precipitation of polymer within the monomer phase after the initiation of polymerisation by means of a monomer-soluble free radical initiator. Interactions between monomer-swollen polymer particles in the monomer phase influence to a large extent the mechanism of formation of a porous polymer network within the grain. The structure of this polymer network affects both the ease with which monomer may be removed from polymer grains and the ability of the product to imbibe plasticiser. Both of these features assume great commercial importance. This chapter reviews work which has led to the development of a model, based on experimental observations, which describes the colloidal phenomena occurring in the polymerising monomer with increasing conversion.

2. EMULSION AND MICROSUSPENSION POLYMERISATION PROCESSES

2.1. Criteria for Design of Particle Size Distributions

The methods of fabrication of dried PVC powders to produce a finished product demand a number of different properties from the polymer. In order to provide these properties the size and size distribution of the primary particles in the polymer grains play an important part. Dried PVC powder may be used either directly in the dry state (e.g. for extrusion applications) or as a plastisol where the PVC powder is dispersed in a plasticiser at weight fractions typically >0·6. Evans[3] has reviewed many aspects of the technology associated with the production of these polymers.

For dry polymer extrusion, where easy polymer gelation is important, powders with a high packing density and good flow properties are required. This requirement is met in practice by using a latex which has a particle diameter in the range 0·02–0·2 μm.

Perhaps the most careful control of latex particle size during polymerisation is needed for the preparation of polymer used for the production of plastisols, where it is necessary to control the plastisol rheology for any given application. Plastisols made from different batches of latex polymer must have a consistently reproducible viscosity at the shear rate relevant to the fabrication process.

Starting with a monodisperse size distribution of primary particles, Palmgren[4] showed that plastisol viscosity decreases with increasing primary particle size, where the effect is most pronounced at low shear rates. For any given particle size, plastisol rheology is also modified by broadening the particle size distribution. The decrease in viscosity obtained in this way is achieved mainly by a reduction of the void space between the particles. A broad distribution in a concentrated dispersion can pack more efficiently than a monodisperse distribution. The generation of a broad distribution of micron-size monomer droplets by homogenisation, followed by polymerisation within the monomer droplets provides the basis for the PVC microsuspension process.

The packing efficiency of a monodisperse latex may be increased by the addition of one or more latices of smaller particle size. Hence certain blends of PVC latices can be used to produce plastisols with low viscosity over a wide range of shear rates.[5] However, it is more usual to produce multidisperse particle size distributions at the polymerisation stage in a seeded emulsion polymerisation process.

2.2. Emulsion Polymerisation
Studies on the emulsion polymerisation of vinyl chloride monomer (VCM) in the absence of emulsifier at low solids content (12–18% w/w)[6] showed that stable monodisperse latices could be produced

and that the particle number concentration, and hence particle size, was consistent with the mechanism of emulsion polymerisation for emulsifier-free systems proposed by Fitch and Tsai.[7,8] This model assumes that each growing radical, produced by homogeneous initiation in the aqueous phase, results in a new particle provided it reaches a sufficiently high degree of polymerisation without being captured by a pre-existing polymer particle. These growing particles may also coagulate until they carry a sufficiently high surface charge to produce colloidally stable latex particles. The number concentration and hence size of the particles have been found to depend on the ionic strength, charge density on the particles and temperature. These observations confirmed that the mechanism of colloidal stability was by electrical double layer repulsion according to the DLVO theory.

Clearly, an objective of commercial production is to obtain a stable latex which has as high a solids content as remains consistent with producing the required balance of polymer properties after drying. The colloidal stability of latices is enhanced by the addition of surfactants. Latices prepared by emulsion polymerisation techniques usually have solids content of 40–45%. The kinetics and mechanism of emulsion polymerisation of VCM have been extensively reviewed by Ugelstad et al.[9]

There are three types of emulsion polymerisation process operated by the PVC industry: batch, semi-batch and continuous. (See Chapter 2 for a more detailed account of these different polymerisation processes.) Continuous processes have limitations in the range of particle size distributions which are obtainable; these constraints are reviewed by Evans[3] and by Clark.[10] In a batch process, where all the ingredients are charged to the reactor at the beginning of polymerisation, the product contains particles which have too small a particle size for most purposes. Initiation of polymer in micelles of an anionic surfactant nucleates a high number density of particles which then remain stable to coagulation. A semi-batch process is more generally operated in which initiator and emulsifier are added continuously throughout polymerisation. Water-soluble initiators such as potassium persulphate and ammonium persulphate are typically used. In order to utilise all the available reactor cooling at the beginning of polymerisation the initiator is often activated by redox catalysis. Typical initiator systems are ammonium persulphate/sodium bisulphite/copper sulphate and hydrogen peroxide/ascorbic acid/ferrous sulphate.[3]

The type and concentration of emulsifier are important in determining particle number concentration and this is particularly important in semi-batch seeded emulsion polymerisation processes. Min and Gostin[11] recently proposed a mathematical model to predict the way in which the final particle size distribution of a typical seeded emulsion polymerisation was generated. The model was tested against pilot-plant experiments using a standard recipe.[12] The polymerisation was seeded using a PVC latex with a particle diameter of approximately $0.5 \mu m$. The emulsifier was a fatty-acid sulphate salt, which together with a water-soluble redox initiator system was injected continuously during polymerisation. At the beginning of the reaction the reactor contains only monomer droplets and monomer-swollen latex particles. After commencement of surfactant injection, the surface coverage of the reactor contents, both seed latex and monomer droplets, by surfactant increases. When the concentration of surfactant is sufficient not only to provide complete coverage on the surface of the reactor contents but also to exceed the critical micelle concentration (CMC) in the aqueous phase, a secondary nucleation of particles initiated in micelles occurs. Secondary nucleation continues provided that the emulsifier concentration in the aqueous phase exceeds the CMC.

The model of Min and Gostin[11] makes a number of assumptions about the course of a seeded emulsion polymerisation. These include: (1) polymer particles are homogeneous with no internal structure; (2) new particles are formed from micelles; (3) polymer particles are stabilised both by surfactant and polymer chain ends at the latex particle surface; (4) latex particles may coagulate; (5) the Tromsdorff effect is present from the early stages of polymerisation; and (6) shrinkage of the reaction volume due to polymerisation is significant and is included in the model. The predicted particle size at different conversions in a seeded emulsion polymerisation using this model is shown in Fig. 1. Micelles appear late in this polymerisation (31% conversion) due to the large amount of seed latex used for this run. The actual particle size distribution at the end of the polymerisation as determined by a Joyce–Loebl disc centrifuge is compared with the prediction of the model in Fig. 2. This shows that seed particles have grown to approximately twice the original diameter together with the generation of another family of particles which have a smaller particle size. The size and ratio of each mode present in this type of seeded emulsion polymerisation may be controlled by variation of seed charge, seed size, initial initiator concentration and the rate of surfactant injection.

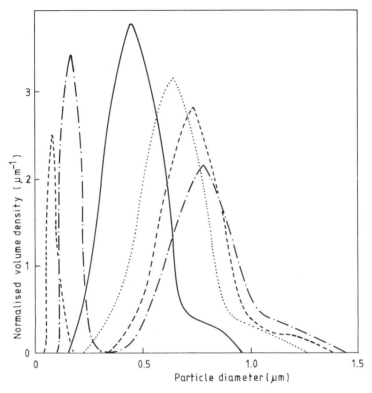

Fig. 1. Predicted particle size distributions in a seeded emulsion polymerisation process at different conversions: ——, 8·5%; · · · ·, 30%; – – – –, 50%; —·—·—, 80%. (Reproduced from Ref. 11 by permission of American Chemical Society.)

2.3. Microsuspension Polymerisation

In this process monomer is homogenised with water, emulsifier and a monomer-soluble initiator. After homogenisation to produce a broad distribution of droplets with modal mean diameter in the region of 1 μm, the concentrated emulsion is pumped to the reactor. The homogenisation process to a large extent determines the final latex particle size in that polymerisation occurs principally within monomer droplets. There is a possibility that some aqueous phase polymerisation occurs by radical transfer from the monomer droplets and that a limited amount of coagulation occurs during the process as a result of

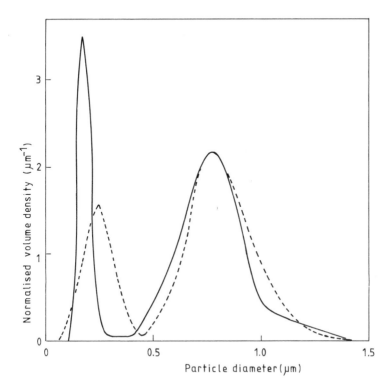

Fig. 2. Predicted and pilot-plant data for the particle size distributions in a PVC seeded emulsion polymerisation process: ——, predicted; – – – –, pilot-plant experiment. (Reproduced from Ref. 11 by permission of American Chemical Society.)

the agitation conditions used. The preferred initiator for this type of polymerisation is a long-chain diacoyl peroxide, particularly dilauroyl peroxide.[3] The particle size distribution produced in this process is very broad, extending from $\approx$0·1 to 1·5 μm, with a few particles of even larger diameter.

Some modifications to the microsuspension process have been developed such as a seeded process[13] and an extension to this where an emulsion polymer latex, particle size $\approx$0·1 μm, is added to the seed latex and monomer at the beginning of the polymerisation.[14] In the latter process it is claimed that by careful choice of the size and concentration of the added emulsion polymer the mechanical stability of the total size distribution of particles remains sufficiently high during polymerisation to allow the production of latices with solids content up to 55% w/w. The viscosity of plastisols obtained using this type of polymer is also claimed to be lower than for plastisols made from similar microsuspension polymer which does not contain the small particles.

2.4. Recent Advances in Polymerisation Techniques

Our understanding of the formation and stability of monomer emulsions which can be subsequently polymerised has been advanced considerably by the work of Ugelstad and co-workers, both in general terms[15] and specifically in relation to PVC.[9] These reviews have suggested several new methods by which PVC latices may be prepared:

(1) Polymerisation of monomer emulsions, formed using mixed emulsifiers.

(2) Two-step swelling methods for producing large monodisperse particles.

In the first of these developments, Ugelstad et al.[16,17] showed that finely dispersed emulsions could be prepared without the use of a homogeniser by using a combination of a long-chain anionic or cationic surfactant together with a long-chain non-ionic surfactant. The first step is to produce a mixture of the surfactants in water (total surfactant concentration 1·5–2·0% w/w) at ≃70°C. At this stage the aqueous phase contains micellar and liquid crystal structures. Following the addition of VCM, 'spontaneous' emulsification occurs, generating a distribution of emulsion droplets in the submicron range. If the emulsifiers are incorporated differently, e.g. if the alcohol is initially dissolved in the monomer phase, a much coarser emulsion results. Ugelstad et al. suggested that the material transfer of long-chain alcohol from micelles to the large monomer droplets may result transiently in a high adsorption of alcohol at the VCM/water interface which then gives rise to spontaneous emulsification.

In conventional emulsion polymerisation using water-soluble initiators, monomer droplets are present only as a reservoir for monomer which must diffuse through the aqueous phase to the locus of polymerisation. The droplets are not thought to participate in the polymerisation because, being relatively large, they have such a low specific surface area that they cannot compete with homogeneous or micellar initiation mechanisms. However, the very fine emulsions generated using a mixed emulsifier system have a sufficiently high specific surface area that initiation within monomer droplets is significant using water-soluble initiators producing latices with a broad particle size distribution.[18]

Using potassium persulphate as the initiator, Ugelstad et al.[9] have demonstrated the effect of different mixed emulsifier combinations on the particle size distribution of PVC polymers. With hexadecanol/sodium dodecyl sulphate (SDS) as the emulsifier combination a large number of fine particles were initiated in the aqueous phase in addition to polymer initiated within monomer droplets. These fine particles were found to produce an adverse effect on latex stability. However using hexadecanol/sodium hexadecyl sulphate (SHS) the latex was free of polymer initiated in the aqueous phase and more colloidally stable. Ugelstad et al. suggested that although the CMC of SDS is higher than SHS, SDS is reluctant to adsorb at the VCM/water interface and is therefore free to stabilise new polymer formed in the aqueous phase. However, the adsorption of SHS at the VCM/water interface is believed to be much greater because of the presence of the hexadecanol.

The two-step swelling process developed by Ugelstad et al.[9,19] is a method by which large monodisperse particles of a predetermined size can be readily prepared. In the first step a low molecular weight/monomer-soluble/water-insoluble component such as chlorododecane or dioctyl adipate was introduced into poly(styrene) seed particles. In order to aid diffusion of the water-insoluble component added to the seed particles as a homogenised emulsion, approximately 10% v/v acetone was added to aqueous phase. The acetone was removed after diffusion of the water-insoluble component to the seed particles was complete. In the second stage VCM was added to the seed latex. The presence of the water-insoluble component in the seed caused the seed particles to swell beyond their equilibrium swelling capacity. Ugelstad et al. proposed that the osmotic pressure difference between the monomer reservoir and the swollen particles caused by the presence of the monomer-soluble/water-insoluble component in the seed completely depleted the reservoir of monomer when equilibrium was reached. In this way, Ugelstad et al. claimed that volume swelling ratios for monomer-swollen seed particles of up to 800:1 are possible. Initiation in these large droplets by monomer-soluble initiators produced PVC particles which were monodisperse in size but not in shape. The non-uniform shape of the particles may indicate that even within monomer droplets of diameter 2–5 μm phase separation of polymer in monomer has occurred, as commonly observed in suspension polymerisation.

Both of these developments together with the work of Ugelstad *et al.*[15] on understanding diffusion processes in emulsion formation should help the PVC industry to achieve a closer control of particle size distribution in their polymerisation processes.

2.5. Polymer Isolation Processes

Typical polymer isolation processes for emulsion and microsuspension polymers have been reviewed by Lovelock.[20] The most common method of obtaining a non-dusty granular powder is to spray-dry a PVC latex which typically has a solids content within the range 35–45% w/w. The processability of the polymer is affected by the particle size of the grains and the extent of fusion of primary particles within the grains. These properties depend on a number of operating parameters particularly the atomisation conditions in the drier and the drier outlet temperature.[20] The grain size distribution is normally broad and in the range 20–50 μm. Electron micrographs of freeze-cut grains of spray-dried PVC latex are shown in Fig. 3. These micrographs illustrate all the elements of the seeded emulsion polymerisation production process. The primary latex particles were generated by a route similar to that described by Min and Gostin[11] and outlined in Section 2.2. The spray-drying process produces spherical grains which often have a hollow core as shown in Fig. 3(a). Two different populations of primary particles can be clearly seen in the sectioned grain at higher magnification (Fig. 3(b)).

Spray-drying is a costly operation requiring a large amount of energy to remove water from the latex. Ways of reducing these drying costs have attracted much research effort. One way is simply to reduce the amount of water in need of removal by spray-drying a latex of higher solids content. Alternatively, grain formation by latex flocculation is also a possibility where most of the water in the resulting slurry may be removed by either filtration or centrifugation. The cost saving which can be made by drying a concentrated latex is shown in Fig. 4. Taking a typical PVC latex with a solids content of 40%, the increase in dry polymer output for the energy input required to remove water from a 40% solids dispersion is plotted against solids content. For the same

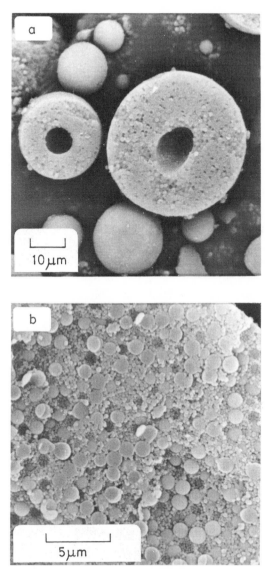

Fig. 3. Freeze-cut sections of PVC grains produced by spray-drying a seeded emulsion polymer latex.

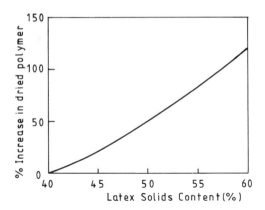

Fig. 4. Increase in dried polymer latex output against latex solids content obtained for the same rate of water removal as from a 40% solids latex.

quantity of water removed an increase in solids content by 10% (from 40 to 50%) increases the dry polymer output by 50%.

After polymerisation it is possible to concentrate the latex prior to spray-drying by either electrodecantation or high-pressure dialysis using semi-permeable membranes.[20] However, it is much more convenient to polymerise to high solids content in the reactor. Polymerisation of PVC to high solids content has been described in the patent literature.[13] However, for emulsion polymers a serious problem is one of mechanical stability during polymerisation. Palmgren[21] has studied the mechanical stability of monodisperse PVC latices as a function of particle size, emulsifier concentration, electrolyte concentration and residual monomer content. The mechanical stability of PVC latices of different particle size containing different surface concentrations of SDS emulsifier is shown in Fig. 5. At a given emulsifier coverage on the latex, these data indicate that the latex stability decreases with increasing particle size. Latex mechanical stability also decreases with both increasing electrolyte concentration and solids content. Methods for maintaining adequate stability of concentrated multi-modal PVC emulsion polymers during polymerisation represent an important area for further research.

As an alternative to spray-drying, Hassander et al.[22] have reported experiments on the granulation of PVC latices using a three-stage process. The stages are (a) addition of a water-soluble polymer to the latex, (b) addition of VCM to the flocculating latex under agitation, (c)

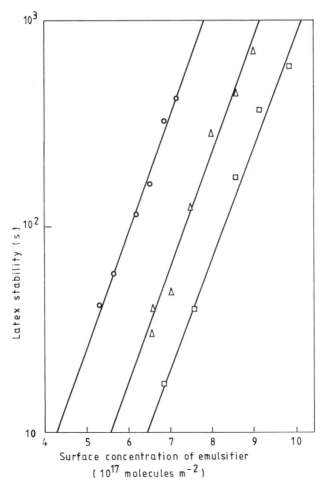

Fig. 5. Mechanical stability of 45% w/w PVC latices as a function of particle size and emulsifier level. Latex particle diameter: ○, 123 nm; △, 202 nm; □, 283 nm. (Reproduced from Ref. 21 by permission of American Chemical Society.)

removal of VCM. A 71·5% hydrolysed polyvinyl acetate (PVAc) was used as the flocculating polymer. Hassander *et al.* showed that the addition of VCM caused flocculated latex to pass into the VCM phase forming macroscopic aggregates, the precursor to the polymer powder. During this process the hydrolysed PVAc was desorbed by the aggregates and passed into the aqueous phase. In removing the VCM, hydrolysed PVAc was readsorbed by the PVC grains while latex particles in the grains became partially fused. This type of flocculation process aids polymer isolation in that the water may be filtered from the grains. Centrifugation of the grains followed by air-drying in the same way as for suspension polymer is then possible.

3. SUSPENSION AND BULK POLYMERISATION PROCESSES

3.1. Generation of Monomer Droplets in Suspension Polymerisation

Unlike styrene and acrylic monomers, VCM is a gas at STP and hence suspensions or emulsions of liquid VCM in water at elevated pressures are not very amenable to study by optical microscopy. A few distributions of VCM suspensions in water viewed by light microscopy into specially designed pressure cells appear in the literature,[23,24] but no analyses of droplet size distribution under different conditions of reactor agitation or polymeric additive addition have been reported. A technique for fixing VCM emulsions by osmium tetroxide[25] may prove useful to study the VCM/water system in greater detail. Mersmann and Grossmann[26] have studied the dispersion of liquids in non-miscible two-phase systems, which include chlorinated liquids such as carbon tetrachloride in water. The influence of stirrer type and speed on the development of an equilibrium droplet size distribution is discussed. Different empirical relationships to calculate the Sauter mean diameter of droplet distributions from reactor operating parameters are also reviewed.

In VCM suspension polymerisation, water and liquid VCM in approximately equal phase volumes are stirred together with a water-soluble polymer such as a partially hydrolysed poly(vinyl acetate) (PVAc) or a substituted cellulose, e.g. hydroxypropyl cellulose, hydroxybutyl methyl cellulose, ethyl hydroxyethyl cellulose. The droplet size distribution obtained is the result of two competing processes, droplet break-up and coalescence; break-up occurs in the turbulent shear field around the impeller tip, coalescence in the regions of the reactor which are distant from the impeller. This process continues until some 'equilibrium' droplet size distribution is obtained; this determines the initial size of polymerising droplets. The initial droplet size distribution depends on (a) stirrer type, (b) stirrer speed, (c) reactor configuration and position of stirrer in the reactor, and (d) nature and concentration of water-soluble polymeric additives.

In general, the more effective the water-soluble polymer is in reducing the interfacial tension between the two phases the smaller the mean size of the droplet distribution. This is because the lowering of interfacial tension promotes droplet break-up, and interfacially adsorbed polymer then protects the droplets against coalescence. The interfacial tension between VCM and water as a function of time at 20°C is shown in Fig. 6 where the aqueous phase contains a low concentration of different macromolecular additives, including a partially hydrolysed PVAc and different types of substituted cellulose. Measurements were made using the spinning drop technique as described by Zichy.[27] Compared with the VCM/water interfacial tension of $31.2 \, \text{mN m}^{-1}$ at 20°C, the data in Fig. 6 demonstrate the large

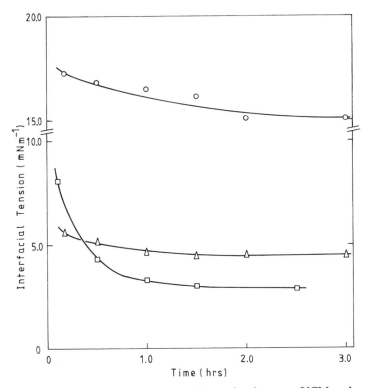

Fig. 6. Time dependence of the interfacial tension between VCM and water containing 10 ppm macromolecular additives at 20°C: ○, hydroxypropyl cellulose; △, hydroxyethyl cellulose; □, partially hydrolysed PVAc.

reduction in interfacial tension which is brought about by the addition of a low concentration of these water-soluble macromolecules. The hydroxypropyl cellulose is more effective in reducing the VCM/water interfacial tension than hydroxyethyl cellulose. The trend for VCM/water interfacial activity is the same as for the surface activity at the air/water interface.[10] The reduction of interfacial tension with time indicates that the polymer molecules adsorbed at the VCM/water interface rearrange their configuration to minimise the interfacial free energy. This effect is most marked for the hydrolysed PVA.

It must be appreciated that small changes in the molecular structure of these macromolecular additives can produce large changes in interfacial adsorption behaviour. Taking partially hydrolysed PVAc as an example, structural variables of the molecule include molecular weight, degree of hydrolysis, sequence length distribution and extent of chain branching. The structure of partially hydrolysed PVAc has been reviewed by Dunn,[28] and the surface and interfacial properties have been studied by Tadros and co-workers[29,30] and by Langveld and Lyklema.[31] This work has concentrated on copolymers which had a degree of hydrolysis of 88%, a material which has been found most effective for the stabilisation of emulsions and is used commercially in the emulsion polymerisation of vinyl acetate.

Hydrolysed PVAc with a degree of hydrolysis in the region 70–80% is more suitable for the suspension polymerisation of VCM. Few studies on materials with these lower degrees of hydrolysis have been reported, although the adsorption properties of these materials at the aqueous solution/air interface have been studied by Zichy et al.[32]

The temperature of polymerisation (50–70°C) is higher than the cloud point in water of hydrolysed PVAc which is typically used. Under these conditions, the hydrolysed PVAc forms a phase-separated polymeric layer at the VCM/water interface although the interfacial tension remains low. At this stage the stability of VCM droplets to coalescence is unlikely to be provided by a steric repulsion of water-soluble poly(vinyl alcohol) loops and tails, but is more likely due to the elastic properties of the interface conferred by the phase-separated hydrolysed PVA. As polymerisation proceeds, the interfacial tension of the monomer droplets increases.[27] The hydrolysed PVAc becomes less effective in stabilising droplets against coalescence as polymer radicals or monomer radicals transfer to the partially hydrolysed PVAc, producing a graft copolymer. This graft copolymer or pericellular membrane around the polymer grains has been identified and isolated by Tregan and Bonnemayre[33] and by Davidson and Witenhafer.[34] Changes in the surface mechanical properties of the polymerising droplet as a result of graft copolymerisation may give rise to a limited coalescence of the droplets at low conversion where each fully polymerised grain contains 5–15 droplets. A higher initial concentration of partially hydrolysed PVAc can prevent coalescence during polymerisation so that one monomer droplet produces one polymer grain. The properties of the two products are different. The multiple-droplet product provides a larger coarser grain which has a higher internal porosity than the single-droplet grain.[35]

3.2. Precipitation of PVC in the Monomer Phase During Polymerisation

Previously published work on the precipitation of polymer gel particles in the monomer phase during bulk and suspension polymerisation is reviewed by Ugelstad et al.[9] More recent contributions relating to parameters which influence the colloidal stability of the precipitated polymer are considered here.

Many different terms are used in the literature to describe the colloidal units which can be detected in the final porous PVC grains by either optical or electron microscopy. The nomenclature which best takes into account colloidal processes which are responsible for the formation of a particulate network in polymer grains was proposed by Allsopp.[36] This nomenclature for sub-units of the grain is shown in Table 1 together with the IUPAC nomenclature.[37] The size of these sub-units and the earliest conversion at which they are formed during

TABLE 1
Sub-units of a PVC Grain

Nomenclature of sub-units appropriate to colloid formation	Particle diameter	Conversion at first appearance of colloidal sub-units	IUPAC nomenclature	Approximate particle diameter
Basic particle	15–20 nm	1×10^{-3}–5×10^{-3}	Microdomain	10 nm
Primary particle nucleus	0·08–0·10 μm	5×10^{-3}	Domain	0·1 μm
Primary particle	0·2–1·0 μm		Primary particle	1 μm
Cluster	2–5 μm	0·02–0·1%	Agglomerate	10 μm
Sub-grain				
Grain			Grain	100 μm

polymerisation have been assigned as a result of more recent experiments.

Colloidal aspects of precipitation polymerisation extend from the conversion at which polymer gel particles phase-separate from the monomer phase to the conversion at which precipitated polymer becomes immobilised as a rigid contiguous network.

3.2.1. Formation of Basic Particles

The nucleation stage of bulk vinyl chloride polymerisation was studied by Boissel and Fischer.[38] From turbidimetric measurements they suggested that polymer precipitated from the monomer phase at conversions as low as 10^{-3}%. Prior to these measurements, estimates of conversion at phase separation were much higher. However, Rance and Zichy[39] suggest that, since the Flory interaction parameter for VCM/PVC $\chi = 0\cdot98$,[40] the solubility of polymer in monomer is limited to polymer chains containing $\simeq 10$ monomer units. This also implies that phase separation occurs at very low conversion. The polymer gel phase is believed to appear as so-called basic particles which could be the result of polymerisation by a single radical produced from the decomposition of an initiator molecule.[39] Each radical produced in this way may by transfer to monomer generate as many as 20 polymer chains. If each chain has $DP \simeq 1000$ and the chains produce a single coil equivalent to a molecule of 20 times the individual chain length, the diameter of such a basic particle has been calculated to be 18 nm.[39] This assumes that the particle comprises polymer gel containing $0\cdot33$ volume fraction monomer.[40]

Direct electron microscopic evidence for basic particles in a polymerising system is scarce, but they have been identified in equilibrium with primary particle nuclei.[41] However, Behrens et al.[42] reported that the smallest particles resolvable by electron microscopy in a grain produced by suspension polymerisation were 10 nm diameter. This view has been supported by Barclay[43] and more recently by Soni et al.[44] who concluded that each basic particle was composed of a crystalline core surrounded by less-ordered material in a likely fringed micelle structure.

3.2.2. Formation and Growth of Primary Particles

The first readily observable particles which precipitate in the monomer phase are the so-called primary particles. Most authors agree on the size of primary particle nuclei ($\approx 0 \cdot 1 \ \mu m$), which seems to be independent of the polymerisation temperature or rate of initiation. Primary particles are formed by the coagulation of basic particles; this appears to be a rapid process after the basic particles are formed. An attempt to study the aggregation of basic particles to form primaries was reported by Rance and Zichy[39] in which the nucleation of polymer in the monomer phase was followed by photon correlation spectroscopy (PCS). Full experimental details are given in Ref. 45.

Bulk polymerisations, initiated by dilauroyl peroxide were carried out over the temperature range 35–60°C. After initiation of polymerisation no evidence for basic particles was observed even for polymerisations with the slowest kinetics. The first particles for which an autocorrelation function could be fitted using the PCS technique were already primary particles with mean diameter $0 \cdot 08 \ \mu m$.

Rance and Zichy[39] described a model for the formation of primary particles from basic particles in which it was assumed that polymer is generated continuously in the early stages of polymerisation as a flux of basic particles. In the absence of any mechanism to confer colloidal stability on the basic particles they coagulate according to the Smoluchowski rate equation. The fate of a freshly formed basic particle is that it may either become the nucleus of a new primary particle or be captured by an existing primary particle. By assuming that the ratio of probabilities of these fates was equal to the ratio of collision cross-sections of basics to primaries, relationships were obtained which described the primary particle size and number as a function of polymerisation time. The model predicted that the equilibrium number of the primary particles rapidly approached a plateau within the first 10–20 s of polymerisation. Since PCS measurements require a sampling time of at least 30 s, it is not surprising that basic particles were not identified by this technique; their separate existence is very short-lived. For this reason, primary particle nuclei are formed at very low conversion, approximately $5 \times 10^{-3}\%$, much earlier than had been previously suggested.

The mechanism of stabilisation for primary particles in the monomer phase over a limited range of conversion has been the subject of much speculation and study. Wilson and Zichy[46] were the first to make direct observations of electrophoresis of precipitated polymer in VCM and to suggest that electrostatic stabilisation was responsible for stability of primary particles. Using a simple moving boundary technique they found that PVC particles in the monomer were negatively charged. A more quantitative approach was taken by Cooper et al.[47] They measured the electrophoretic mobility of precipitated polymer diluted by liquid VCM in a quartz microelectrophoresis cell designed to operate at the autogenous pressure of VCM (up to 10 bar). They established that primary particles underwent electrophoresis and measured an electrophoretic mobility of $-1 \cdot 2 \times 10^{-8} \, \mathrm{m^2 \, V^{-1} \, s^{-1}}$ for a dispersion of $0 \cdot 3 \, \mu \mathrm{m}$ diameter PVC particles in liquid VCM. This corresponds to a ζ potential of $-83 \, \mathrm{mV}$. Davidson and Witenhafer[34] have also measured electrophoretic mobility of nascent PVC particles in monomer, using a moving boundary technique, and report values for ζ potential up to approximately $-120 \, \mathrm{mV}$, although no particle size data were provided.

Taking the data of Cooper et al., Rance and Zichy[48] compared the number of charges at the plane of shear per PVC particle in VCM with the surface charge of typical aqueous anionogenic polymer latices of similar particle size. These data are shown in Table 2. Values of Lifshitz–Hamaker constants for these three systems indicate that the attractive interactions between particles are very similar. The data in Table 2 show that colloidal stability of PVC particles in VCM is achieved at a very much lower surface charge density than for typical aqueous latices. This may be understood with reference to Fig. 7 which shows the total potential energy of interaction $V_T/\mathbf{k}T$ against distance of separation between two PVC particles in VCM (Curve A) and between poly(styrene) particles in an aqueous medium (Curve B).[48] The total potential energy of interaction between two particles was calculated as the sum of contributions from the electrostatic potential energy and van der Waals attractive potential energy according to the DLVO theory. Approximate expressions for these contributions appropriate to interaction between spheres are given in Chapter 4. Values of V_T were calculated assuming that in both cases the particles

TABLE 2
Number of Charges per Particle and Lifshitz–Hamaker Constants for Different Polymer Latex Systems

Latex system	Charges/particle at plane of shear	Charges/ particle at surface	A_L $(10^{-21} \, J)$
PVC/VCM	41	—	5·0
PTFE/water	—	$5 \cdot 7 \times 10^3$	3·6
Poly(styrene)/water	—	$7 \cdot 1 \times 10^4$	9·0

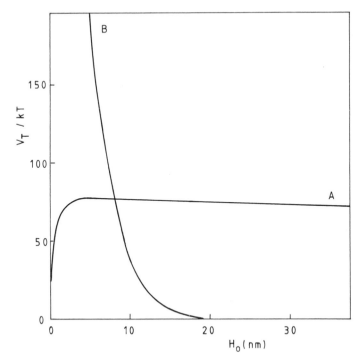

Fig. 7. Diagram of total potential energy (V_T) against distance of interparticle separation (H_0) for spherical particles of radius $0\cdot15\,\mu$m and ζ potential $-80\,$mV. (A) PVC particles in VCM; (B) poly(styrene) particles in $10^{-2}\,$mol dm^{-3} 1:1 electrolyte solution. (Reproduced from Ref. 48 by permission of IPC Business Press.)

had a radius $0\cdot15\,\mu$m and ζ potential $-80\,$mV, and that the dispersion medium for poly(styrene) particles was $10^{-2}\,$mol dm^{-3} 1:1 electrolyte. Figure 7 shows that the potential energy maximum of Curve A is much smaller than that of Curve B, reflecting the lower dielectric constant VCM compared with an aqueous solution. However, with increasing distance of interparticle separation, Curve A decays much more slowly than Curve B indicating that the electrical double layer surrounding a particle in VCM $(1/\kappa > 10\,\mu$m) is much thicker than in an aqueous medium $(1/\kappa = 3\,$nm).

Although PVC primary particles dispersed in VCM have been shown to undergo electrophoresis, the origin of the charge associated with the particles is still the subject of speculation. Both Rance and

Zichy[48] and Davidson and Witenhafer[34] have suggested that the most likely source of negatively charged ions in VCM at the beginning of polymerisation is chloride ions which may specifically adsorb at the VCM/PVC interface. The chloride ions may be produced either by the dehydrochlorination of PVC or as the product of decomposition of VCM peroxide formed by the reaction of residual oxygen with VCM after monomer charging. An alternative explanation for the negative charge associated with a non-ionogenic dispersion has been recently proposed by Fowkes[49] in terms of electron injection from the dispersion medium into the polymer. However, any proposed charging mechanism is difficult to prove experimentally for this system.

3.2.3. Aggregation of Primary Particles and the Generation of a Continuous Polymer Network within Grains

While primary particles are stable for a long time if the polymerisation is stopped at an early stage in the polymerisation, they do not remain kinetically stable as polymerisation proceeds. Willmouth et al.[45] have used PCS together with simultaneous light transmittance measurements to calculate both the size and number density of PVC particles during the initial stages of quiescent bulk VCM polymerisation. Data for particle size, particle number density and overall volume fraction of nascent PVC gel particles generated during polymerisation at 35°C using 0·1% dilauroyl peroxide as initiator are shown in Fig. 8. The plot of particle size against time consists of two distinct regions, both showing a linear increase in diameter with time and separated by a clear-cut transition. After this transition, there is also an approximately linear decrease in particle number density with time. The transition is believed to represent the onset of aggregation of hitherto stable primary particles. Speirs[41] suggested that the very sharp onset of flocculation of primary particles in the polymerising system is characteristic of a system which contains a constant total charge. The formation of clusters of primary particles commences in a quiescent system at conversions as low as 0·02–0·1%.

The effect of shear on the stability of primary particles and aggregates has been studied in mass polymerisation by Boissel and Fischer[38] and during suspension polymerisation by Davidson and Witenhafer.[34] While the monomer phase experiences high shear fields directly at the stirrer tip during mass polymerisation, the shear field experienced by monomer droplets in suspension polymerisation is also high. In the early stages of polymerisation this has been shown[34] to influence the

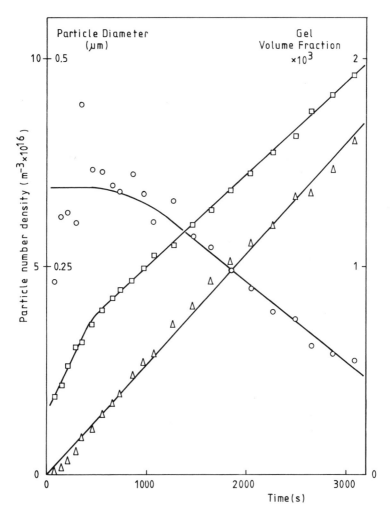

Fig. 8. Data from light-scattering measurements on monomer-swollen PVC particles in VCM as a function of polymerisation time at 35°C: O, particle number density; □, particle diameter; △, gel volume fraction.

colloidal stability of polymer particles precipitated in the monomer phase. Hence in a stirred reactor the colloidal stability of precipitated polymer in the monomer phase for both processes is controlled not only by electrostatic interactions but also by the input of mechanical energy into the system.

Flocculation and growth of clusters within monomer droplets continue until a critical conversion at which the polymer aggregates touch and become immobilised within the droplet as a continuous particulate network. After this stage, polymer sub-units of colloidal dimensions no longer exist; this stage is represented by the broken line in Table 1. Further polymerisation continues both from the monomer phase onto the polymer network and within the polymer gel. The electron micrograph in Fig. 9 is typical of the particulate structure which might be found in a PVC grain at intermediate conversion, although it is a section of polymer produced in a spinning drop polymerisation to 28% conversion.[32] The clusters of primary particles, approximately 2 μm diameter, can be readily identified. The two-phase system ceases to exist after approximately 70% conversion when no free liquid monomer remains. The polymerisation is finally stopped when the reactor pressure reaches a predetermined pressure below autogenous pressure, corresponding to a conversion in the range 80–95%.

Resulting from a number of studies of bulk and suspension polymerisation at different conversions, Rance and Zichy[39] have constructed a consistent sequence of events during polymerisation leading from the liquid monomer through a colloidal two-phase system to the porous solid polymer. This is shown in Fig. 10; polymerisation occurs from left

Fig. 9. Freeze-fracture section of PVC precipitated within a monomer droplet which was inertially suspended in water in a rotating tube (polymerisation terminated at 28% conversion). (Reproduced from Ref. 27 by permission of Marcel Dekker.)

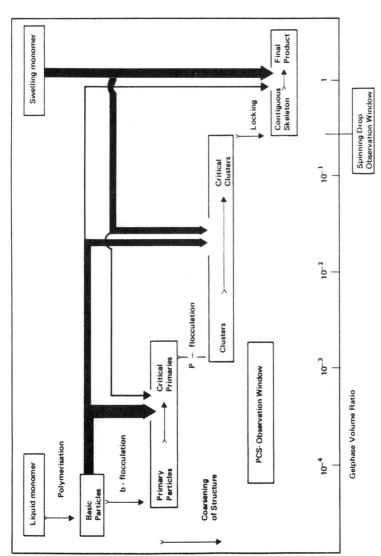

Fig. 10. The life cycle of the two-phase system in vinyl chloride polymerisation. (Reproduced from Ref. 39 by permission of IUPAC.)

to right and the relative importance of accretion of polymer from monomer phase and gel phase polymerisation is indicated by the thickness of the arrows.

4. SUMMARY

The careful control of colloidal events which occur during commercial PVC polymerisation processes is important for obtaining products having consistent processing properties. Microsuspension and emulsion polymerisation processes produce latices in the micron range. Colloidal aspects of the production of the materials include an understanding of the factors controlling the generation of emulsion droplet size distributions, the development of particle size distributions during seeded emulsion polymerisation, and the control of latex stability during polymerisation to high solids content. Polymer isolation processes are also discussed; these may also confer different properties to the final polymer. In the suspension polymerisation process where monomer droplets are suspended in the aqueous phase prior to polymerisation within droplets, factors affecting the mean droplet size and droplet size distribution are considered, particularly the nature of interactions at the VCM/water interface caused by the adsorption of an amphipathic water-soluble polymer. Common to both the bulk and suspension processes is the initiation of polymerisation in the monomer phase. Since PVC is insoluble in its monomer, the precipitation of polymer gel particles in the monomer followed by multiple aggregation and growth of these particles produces porous polymer grains. The characterisation of colloidal particles generated in the early stages of polymerisation is reviewed, together with parameters which influence the stability of precipitated polymer up to the conversion at which a rigid particulate network is formed.

ACKNOWLEDGEMENTS

The author acknowledges the assistance of colleagues at ICI and would particularly like to thank Mr A. Cobbold for electron microscopy, Mr J. C. Wilson for experimental assistance, and Drs R. H. Burgess, A. Frangou and P. V. Smallwood for their constructive criticism of the manuscript.

REFERENCES

1. Burgess, R. H. (ed.), *Manufacturing and processing of PVC*, Applied Science Publishers Ltd, London (1982).
2. Butters, G. (ed.), *Particulate nature of PVC: formation structure and processing*, Applied Science Publishers Ltd, London (1982).
3. Evans, D. E. M. E., In *Manufacturing and processing of PVC*, R. H. Burgess (ed.), Applied Science Publishers Ltd, London (1982), p. 63.
4. Palmgren, O., *Proceedings international symposium on macromolecules*, Helsinki, IUPAC, London (1972).
5. Rangnes, P. and Palmgren, O., *J. Polymer Sci.*, Part C, **33,** 181 (1971).
6. Laaksonen, J. and Stenius, P., *Plastics and Rubber: Materials and Applications*, **5,** 21 (1980).
7. Fitch, R. M. and Tsai, C. H., *J. Polymer Sci.*, Part B, **8,** 703 (1970).
8. Fitch, R. M. and Tsai, C. H., In *Polymer colloids*, R. M. Fitch (ed.), Plenum Press, New York (1971), p. 73.
9. Ugelstad, J., Mork, P. C., Hansen, F. K., Kaggerud, K. H. and Ellingsen, T., *Pure Appl. Chem.*, **53,** 323 (1981).
10. Clark, M., In *Particulate nature of PVC: formation, structure and processing*, G. Butters (ed.), Applied Science Publishers Ltd, London (1982), p. 1.
11. Min, K. W. and Gostin, H. I., *Ind. Eng. Chem. Prod. Res. Dev.*, **18**(4), 272 (1979).
12. Jones, W. D. and Schaefer, S. W., US Patent 3317495 (1967).
13. Fischer, N., Boissel, J., Kemp, T. and Eyer, H., British Patent 1435425 (1976).
14. Kemp, T., British Patent 1503247 (1978).
15. Ugelstad, J., Mork, P. C., Kaggerud, K. H., Ellingsen, T. and Berge, A., *Adv. Colloid Interface Sci.*, **13,** 101 (1980).
16. Ugelstad, J., El-Aasser, M. S. and Vanderhoff, J. W., *J. Polymer Sci.*, *Polymer Letters Ed.*, **11,** 503 (1973).
17. Azad, A. R. M., Ugelstad, J., Fitch, R. M. and Hansen, F. K., *ACS Symposium Series, Emulsion Polymerisation*, **24,** 1 (1976).
18. Kema Nord AB, German Patent 2629655 (1976).
19. Ugelstad, J., Kaggerud, K. H., Hansen, F. K. and Berge, A., *Makromol. Chem.*, **180,** 373 (1979).
20. Lovelock, V. G., In *Manufacturing and processing of PVC*, R. H. Burgess (ed.), Applied Science Publishers Ltd, London (1982), p. 123.
21. Palmgren, O., *ACS Symposium Series, Emulsion Polymerisation*, **24,** 258 (1976).
22. Hassander, H., Nilsson, H., Silvegren, C. and Tornell, B., In *Polymer colloids II*, R. M. Fitch (ed.), Plenum Press, New York (1980), p. 511.
23. Ueda, T., Takeuchi, K. and Kato, M., *J. Polymer Sci.*, *Polymer Chem. Ed.*, **10,** 2841 (1972).
24. Sanderson, A. K., *British Polymer J.*, **12,** 186 (1980).
25. Yang, H. W. H., Bush, C. N., Poledna, D. J. and Smith, R. W., Abstracts 3rd Int. Symposium on PVC, Cleveland, Ohio (1980), p. 135.
26. Mersmann, A. and Grossmann, H., *Chem.-Ing.-Tech.*, **52,** 621 (1980).

27. Zichy, E. L., *J. Macromol. Sci.-Chem.*, **A11**(7), 1205 (1977).
28. Dunn, A. S., *Chem. Ind.*, 801 (1980).
29. Boomgaard, Th. van den, King, T. A., Tadros, Th. F., Tang, H. and Vincent, B., *J. Colloid Interface Sci.*, **66**, 68 (1978).
30. Lambe, R., Tadros, Th. F. and Vincent, B., *J. Colloid Interface Sci.*, **66**, 77 (1978).
31. Langveld, J. M. G. and Lyklema, J., *J. Colloid Interface Sci.*, **41**, 454 (1972); ibid, **41**, 466 (1972); ibid, **41**, 475 (1972).
32. Zichy, E. L., Morley, J. G. and Rodriguez, F., In *Chemie, Physikalische Chemie und Anwendungstechnik der grenzflachenaktiven Stoffe*, Carl Hanser Verlag, Munich (1973), p. 241.
33. Tregan, R. and Bonnemayre, A., *Plast. Mod. Elastomeres*, **23**, 220 (1971).
34. Davidson, J. A. and Witenhafer, D. E., *J. Polymer Sci., Polym. Phys. Ed.*, **18**, 51 (1980).
35. Allsopp, M. W., *Pure Appl. Chem.*, **53**, 449 (1981).
36. Allsopp, M. W., *J. Macromol, Sci.-Chem.*, **A11**(7), 1223 (1977).
37. Geil, P. H., *J. Macromol, Sci.-Phys.*, **B14**(1), 171 (1977).
38. Boissel, J. and Fischer, N., *J. Macromol, Sci.-Chem.*, **A11**(7), 1249 (1977).
39. Rance, D. G. and Zichy, E. L., *Pure Appl. Chem.*, **53**, 377 (1981).
40. Berens, A. L., *Polymer*, **18**, 697 (1977).
41. Speirs, R. M., PhD Thesis, University of Edinburgh (1980).
42. Behrens, H., Griebel, G., Meinel, L., Reichenbach, H., Schulze, G., Schenk, W. and Walter, K., *Plaste Kautsch.*, **22**, 414 (1975).
43. Barclay, L. M., *Angew Makromol. Chemie*, **52**, 1 (1976).
44. Soni, P. L., Collins, E. A. and Geil, P. H., Abstracts 3rd Int. Symposium on PVC, Cleveland, Ohio (1980), p. 105.
45. Willmouth, F. M., Rance, D. G. and Henman, K. M., *Polymer*, **25**, 1185 (1984).
46. Wilson, J. C. and Zichy, E. L., *Polymer*, **20**, 264 (1979).
47. Cooper, W. D., Speirs, R. M., Wilson, J. C. and Zichy, E. L., *Polymer*, **20**, 265 (1979).
48. Rance, D. G. and Zichy, E. L., *Polymer*, **20**, 266 (1979).
49. Fowkes, F. M., *Org. Coat Plast. Chem.*, **42**, 169 (1980).

Index